MATSUTAKE WORLDS

Studies in Social Analysis

Editors: Judith Bovensiepen, *University of Kent*
Martin Holbraad, *University College London*
Hans Steinmüller, *London School of Economics*

Focusing on analysis as a meeting ground of the empirical and the conceptual, this series provides a platform for exploring anthropological approaches to social analysis while seeking to open new avenues of communication between anthropology and the humanities, as well as other social sciences.

Volume 1
Being Godless: Ethnographies of Atheism and Non-Religion
Edited by Ruy Llera Blanes and Galina Oustinova-Stjepanovic

Volume 2
Emptiness and Fullness: Ethnographies of Lack and Desire in Contemporary China
Edited by Susanne Bregnbæk and Mikkel Bunkenborg

Volume 3
Straying from the Straight Path: How Senses of Failure Invigorate Lived Religion
Edited by David Kloos and Daan Beekers

Volume 4
Stategraphy: Toward a Relational Anthropology of the State
Edited by Tatjana Thelen, Larissa Vetters, and Keebet von Benda-Beckmann

Volume 5
Affective States: Entanglements, Suspensions, Suspicions
Edited by Mateusz Laszczkowski and Madeleine Reeves

Volume 6
Animism beyond the Soul: Ontology, Reflexivity, and the Making of Anthropological Knowledge
Edited by Katherine Swancutt and Mireille Mazard

Volume 7
Hierarchy and Value: Comparative Perspectives on Moral Order
Edited by Jason Hickel and Naomi Haynes
Afterword by David Graeber

Volume 8
Post-Ottoman Topologies: The Presence of the Past in the Era of the Nation-State
Edited by Nicolas Argenti

Volume 9
Multiple Nature-Cultures, Diverse Anthropologies
Edited by Casper Bruun Jensen and Atsuro Morita

Volume 10
Money Counts: Revisiting Economic Calculation
Edited by Mario Schmidt and Sandy Ross

Volume 11
States of Imitation: Mimetic Governmentality and Colonial Rule
Edited by Patrice Ladwig and Ricardo Roque

Volume 12
Matsutake Worlds
Edited by Lieba Faier and Michael J. Hathaway

MATSUTAKE WORLDS

Edited by

Lieba Faier and Michael J. Hathaway

berghahn

NEW YORK · OXFORD

www.berghahnbooks.com

First published in 2021 by

Berghahn Books

www.berghahnbooks.com

© 2021 Berghahn Books

Originally published as a special issue of *Social Analysis*, volume 62, issue 4.

Library of Congress Cataloging-in-Publication Data

Names: Faier, Lieba, editor. | Hathaway, Michael J., editor.
Title: Matsutake worlds / edited by Lieba Faier and Michael J. Hathaway.
Description: New York : Berghahn, 2021. | Series: Studies in social
    analysis ; volume 12 | "Originally published as a special issue of
    Social Analysis, volume 62, issue 4"–Colophon. | Includes
    bibliographical references and index.
Identifiers: LCCN 2021015648 | ISBN 9781800730960 (hardback) |
    ISBN 9781800730977 (paperback) | ISBN 9781800730984 (ebook)
Subjects: LCSH: Tricholoma matsutake. | Rare fungi.
Classification: LCC QK629.T73 M38 2021 | DDC 579.5/16–dc23
LC record available at https://lccn.loc.gov/2021015648

British Library Cataloguing in Publication Data

A catalogue record for this book is available from the British Library.

CONTENTS

ILLUSTRATIONS

Figures

Table

FOREWORD

What should the foreword to a mycelial mass look like? Unfortunately for narrative coherency and widely held prose conventions, it should go in multiple directions at once. Keeping with the general role that fungi are understood to play in ecosystems, it should also decompose as it is read. Or, conversely, it should *synthesize* novel materials. In that spirit, here are two interchangeable thematic lenses, a kind of folding pocket magnifier one might usefully stash in one's pocket on the way into a set of matsutake ethnographies.

One. A fungal turn takes everything with it. This book and the world it is being launched into are going through something of a fungal turn. In these days of planetary health concerns and the search for ways to grow food without synthetic fertilizer and pesticide inputs, it is quite fashionable to speak of 'ecosystem services' (Robinson et al. 2017) provided by fungi to the soil, plants, animals, and humans, due to their ability to digest rock and derive minerals and to traffic nutrients and water underground. Such language implies we are expecting the mushroom bill to arrive: $37 squillion for mineral provision and support of 80 percent of the world's terrestrial plant species–and the cost of detoxification is going way up. Of course, many foresters and agronomists will tell you that the bill has already arrived in the great suffering woodlands and expanding deserts as well as the globe's farmlands, which are showing the unexpected wages of tillage, intensive extractive agriculture, and soil fumigants. In many ways the ethnographic accounts found here, each of which attends a different scene of matsutake knowledge making, joins a paradoxical profusion of realization of the depth and importance of interconnection—at the site of its fundamental rupture.

But what does it mean to turn to fungi? Like others of its kingdom, matsutake cannot be taken as a subject without attending to all the other bacterial, plant, invertebrate, and human animal associates that live with, on, in, and in concert with it. As philosophers of biology John Dupré and Maureen O'Malley (2009) might phrase it, the relevant unit of analysis, or selection, or adaptation is not a single replicating lineage taxonomically delineated as a species, nor a metabolically self-sustaining whole that enters into economic transactions with other species. Instead, it is a combination: "Life … is typically found at the collaborative intersections of many lineages" (ibid: 2). Organisms that conventionally would be designated separate species are only able to continue over time when enmeshed in highly specific metabolic relations of cooperation and competition with others. Multi-lineage composite entities collaborating in metabolism are in this view the reproducing biological individuals on which selection acts, and which provide the most satisfying definition to the perennial question of what life is. Such is 'mushroom individuality' (Molter 2017).

Philosophers of biology, of course, are not inclined to incorporate any nod to society or culture in their unending pursuit of better definitions of life and property-cluster kinds. However, we can complement these insights with the stark and urgent terms in which soil conservationists, food sovereignty advocates, and many other figures in this book are experiencing the loss of such long-standing 'collaborative intersections' of organisms with one another and with humans (Baker 2020; Hoover 2018; Puig de la Bellacasa 2019). It is not just food that is at stake when the PCB-polluted fish are banned from consumption, the dam is built, or the mushroom becomes radioactive—or is gone. Rather, the eating-reproducing temporal fabric of biology and culture, and the capacity of such collaborations for 'collective continuance', as philosopher Kyle Whyte (2018) phrases it, disappears. As scholars, we cannot turn analytically or methodologically to animals, then insects, then plants, then bacteria, and wonder if archaea are next. A fungal turn takes everything with it.

Two. The pair term for anthropogenic would be mycogenic. In the pages that follow, we learn that matsutake is not a natural thing that becomes anthropogenically altered with the massive creep of human activity into all aspects of geology and nature—the mark of the Anthropocene—but an event that has anthropogenic elements already multifariously folded into its long evolutionary past. More recently, the socio-economic history of propane displacing wood burning and woodland clearing becomes the historical biology of nematodes and pines and fungi, becomes the market price or the cheap soup substitute meant to simulate the matsutake experience, becomes language or its loss, becomes restoration movements and knowledge about symbiosis that in turn forms the basis of the landscape coordination that becomes matsutake, inducing euphoria.

How to think about the anthropogenic element here? If one were to squeeze this book, its exudate would be a broth of multispecies entanglement thinking. On the one hand, this makes sense in terms of the multispecies rubric as it has been widely used to date: to decenter the human, to knock it down a few pegs, and to make humans a bit part of the ecology rather than its master (Kirksey and Helmreich 2010). On the other hand, the essays interestingly point beyond the conclusions they themselves come to. The human buys, gifts, sniffs, and eats matsutake in the following pages, but the mushroom-in-the-body is oddly missing. Yet if the fungal turn here takes everything with it, it takes humans too.

The hyphae lead in all directions, and thus, paradoxically, we need to go into the human body to follow the non-human. If matsutake are in part an anthropogenic form, then humans are in part a mycogenic form. As noted above, fungi are a bit straitjacketed by their reputation as decomposers, probably because of this business of extruding enzymes and then sucking the digest back in. We are not particularly at ease with life forms that do their eating outside themselves, despite the fact that most fungi do not prefer live humans as substrate due to their uncomfortably high temperatures. But working with all this substrate, fungi are simply amazing synthesizers and make many biochemical entities, both liquid and gaseous, that are found nowhere else.

To illustrate, one molecule synthesized prodigiously in matsutake metabolism is called ergothioneine, an unusual sulphur-containing amino acid constructed only by non-yeast fungi and some soil-dwelling bacteria (Borodina et al. 2020). Rather dismissively called a 'secondary metabolite'—because it seems to be useful to the organism that makes it but has up until now been judged by human scientists not essential to its growth, development, or reproduction—ergothioneine has remained since its discovery in ergot in 1909 something of a residual category, a second thought of unknown utility. Yet plants and animals, including humans, take up ergothioneine with great avidity through a specific membrane transporter that conveys it into cells (Gründemann et al. 2005). Plants such as oats and beans that live in symbiotic relation with mycorrhizal fungi also accumulate ergothioneine, but it is mushrooms that are human's primary dietary source of the molecule.

That plants, animals, and humans are enzymatically mute and do not themselves possess the capacity to synthesize ergothioneine, but harbor a gene that expresses a protein whose specific role is to transport these mushroom-derived molecules into cellular interiors and keep them there, suggests that this is a close and enduring collaboration. Not only do animal cells express a transport protein specific to ergothioneine, but its expression appears to be upregulated chronobiologically prior to mealtimes, which puts rather a new slant on the idea of hunger and appetite (Akamine et al. 2015). Best guess so far is that ergothioneine is useful to cells as an antioxidant, water-soluble and highly stable, able to donate electrons to mute the damaging effects of oxidants. Oxidants are very unstable and reactive due to their unpaired electrons; they arise constantly in cells from exposure to UV light or heavy metals, or are generated by mitochondria as a byproduct of metabolism. Antioxidants help cells with and through such dangers that are part of metabolic living in a world of oxygen (Lane 2002). Perhaps mitochondria, who long ago were free-living bacteria, express a memory of their ancient fungal partners by exploiting such mushroom-derived help (Picard and Burelle 2012).

Ergothioneine is taken up by human cells, but it is not uniformly present in the same quantities everywhere in a body. It is found in highest concentrations in tissues such as bone marrow, the lens of the eye, and seminal fluid—tissues that experience a high amount of oxidative stress (Borodina et al. 2020). This has led to some debate about whether this molecule is a "possible vitamin" that has the paradoxical qualities of being both very widespread in food and yet, under current Western dietary conditions, in "functional undersupply" (ibid.: 190), and is thus failing in its long-standing role as "cytoprotectant" (ibid.: 198). Modern agricultural techniques, conducted with absolutely no eye to mycelial-plant partnerships, have introduced a kind of ergothioneine starvation into modern diets, representative of a more general dynamic in which as the biodiversity of both crops and the soil that they are grown in diminishes, so does human micronutrient consumption (Beelman et al. 2020). Although savvy marketers are already descending to sell matsutake ergothioneine in face creams and longevity supplements—generating yet another commodity form for matsutake—and this is only one of many interesting molecules made in the microbial world and accumulated by plants and animals, the argument still holds. That this fungal metabolite should participate in human sight and reproduction points to aspects by which we might understand the mycogenic nature of humans and other animals, and the consequences of its loss.

As this volume shows, one can easily get carried away with matsutake, but this is probably enough in the way of a decompositional/synthesizing foreword. You used to go down rabbit holes, or profess that your methods tended to the rhizomatic, but after reading this set of works about fungi and its relations, you may instead prefer to descend into hyphal tunnels. No offense to rabbits or roots, but the possibilities opened up as mycelial tips excrete organic acids to burrow through the feldspars and hornblendes under forests, build networks and mantles around and between plant roots, synthesize biochemicals found nowhere else in the living world, and engage in complex parasexual tactics of nucleus swapping ... well, the potential is indeed vast (Hoffland et al. 2003; Molter 2017). Eat up.

Hannah Landecker holds a joint appointment in the Department of Sociology and the Institute for Society and Genetics at the University of California Los Angeles. She is a historian and sociologist of the modern life sciences and biotechnology, with a particular focus on cell biology, metabolism, and epigenetics. Her publications include *Culturing Life: How Cells Became Technologies* (2007), and many articles on the use of film technology in the biosciences, the rise of antibiotic resistance, and the history and sociology of metabolism and metabolic disorders.

References

Akamine, Takahiro, Satoru Koyanagi, Naoki Kusunose, et al. 2015. "Dosing Time-Dependent Changes in the Analgesic Effect of Pregabalin on Diabetic Neuropathy in Mice." *Journal of Pharmacology and Experimental Therapeutics* 354 (1): 65–72. https://doi.org/10.1124/jpet.115.223891.

Baker, Janelle Marie. 2020. "Do Berries Listen? Berries as Indicators, Ancestors, and Agents in Canada's Oil Sands Region." *Ethnos* 86 (1): 1–22.

Beelman, Robert B., Michael D. Kalaras, Allen T. Phillips, and John P. Richie Jr. 2020. "Is Ergothioneine a 'Longevity Vitamin' Limited in the American Diet?" *Journal of Nutritional Science* 9. https://doi.org/10.1017/jns.2020.44.

Borodina, Irina, Louise C. Kenny, Cathal M. McCarthy, et al. 2020. "The Biology of Ergothioneine, an Antioxidant Nutraceutical." *Nutrition Research Reviews* 33 (2): 190–217. https://doi.org/10.1017/S0954422419000301.

Dupré, John, and Maureen A. O'Malley. 2009. "Varieties of Living Things: Life at the Intersection of Lineage and Metabolism." *Philosophy and Theory in Biology* 1 (3): 1–25. https://doi.org/10.3998/ptb.6959004.0001.003.

Gründemann, Dirk, Stephanie Harlfinger, Stefan Golz, et al. 2005. "Discovery of the Ergothioneine Transporter." *Proceedings of the National Academy of Sciences* 102 (14): 5256–5261. https://doi.org/10.1073/pnas.0408624102.

Hoffland, Ellis, Reiner Giesler, Antoine G. Jongmans, and Nico van Breemen. 2003. "Feldspar Tunneling by Fungi along Natural Productivity Gradients." *Ecosystems* 6 (8): 739–746.

Hoover, Elizabeth. 2018. "Environmental Reproductive Justice: Intersections in an American Indian Community Impacted by Environmental Contamination." *Environmental Sociology* 4 (1): 8–21.

Kirksey, S. Eben, and Stefan Helmreich. 2010. "The Emergence of Multispecies Ethnography." *Cultural Anthropology* 25 (4): 545–576.

Lane, Nick. 2002. *Oxygen: The Molecule That Made the World*. New York: Oxford University Press.

Molter, Dan. 2017. "On Mushroom Individuality." *Philosophy of Science* 84 (5): 1117–1127. https://doi.org/10.1086/694011.

Picard, Martin, and Yan Burelle. 2012. "Mitochondria: Starving to Reach Quorum?" *BioEssays* 34 (4): 272–274. https://doi.org/10.1002/bies.201100179.

Puig de la Bellacasa, Maria. 2019. "Re-animating Soils: Transforming Human-Soil Affections through Science, Culture and Community." *Sociological Review* 67 (2): 391–407. https://doi.org/10.1177/0038026119830601.

Robinson, David A., Fiona Seaton, Katrina Sharps, et al. 2017. "Soil Resources, the Delivery of Ecosystem Services and Value." *Oxford Research Encyclopedia of Environmental Science*, 20 November. https://doi.org/10.1093/acrefore/9780199389414.013.375.

Whyte, Kyle Powys. 2018. "Food Sovereignty, Justice, and Indigenous Peoples: An Essay on Settler Colonialism and Collective Continuance." In *The Oxford Handbook of Food Ethics*, ed. Anne Barnhill, Mark Budolfson, and Tyler Doggett, 345–366. New York: Oxford University Press.

INTRODUCTION
Elusive Matsutake

Lieba Faier for the Matsutake Worlds Research Group

This book explores the multispecies relationships of transnational matsutake worlds. It does so in two ways. First, it focuses on multispecies engagements to cast new light on transnational connections. At the same time, it asks what a mushroom's cosmopolitan itineraries can teach us about the dynamics of multispecies worlds.[1] Thus far, studies of transnational processes have tended to center on human activities (Appadurai 1996; Clarke 2004; Inda and Rosaldo 2008; Kearney 1995; Massey 1994; Ong 1999, 2003, 2006; Trouillot 2003). Even when these studies reference more-than-human beings, they often center their analyses on movements of people, capital, information, or technology (Lowe 2010; Mitchell 2002; Nading 2017).

In contrast, we draw attention to multispecies encounters as constitutive of transnational processes. We explore the interconnected, cosmopolitan lives of humans, mushrooms, trees, insects, and other creatures. We argue that to

Notes for this section begin on page 12.

understand these multispecies relationships we need to pay attention to their various dynamics of 'elusiveness'. As explained below, we use this term to mark the productive coordinations that develop in encounters among diverse ways of being, despite the impossibilities of direct translation between them. To track these dynamics across geographical and species lines, we turn to collaboration. Matsutake are our guides.

Matsutake are the fruiting bodies of ectomycorrhizal fungi, fungi that live mutualistically with the roots of certain species of trees. They are eaten as a gourmet delicacy, primarily in Japan. Humans find the mushrooms elusive for a variety of reasons. The matsutake fungus is difficult to pin down geographically, having been discovered with a range of different hosts in sites as far-flung as Turkey, Korea, Canada, Japan, China, Norway, and Mexico (Wang et al. 1997). The main host of matsutake fungi in Japan is the red pine (*Pinus densiflora*). However, they have also been found in relationships with other kinds of pines as well as with some oaks, firs, spruce, cedar, chinkapin, and other trees. Despite considerable efforts by scientists, matsutake have thus far eluded humans' attempts at their cultivation. They fruit fleetingly and unpredictably, evading human plans for control and instead popping up in relation to their own life-world agendas. They are also difficult to find. Hidden in the forest duff, they are often invisible to the untrained eye. The challenges of procuring the mushrooms have made them prohibitively expensive for most consumers, and even those who love the mushroom find themselves at a loss for words when asked to characterize its aromatic appeal (Inoue, this volume). Moreover, because matsutake growth depends on fungal relationships with trees and other forest dwellers, scientists and would-be cultivators have difficulty conceptually isolating the mushrooms from the web of natural-social relationships through which they grow. Some Japanese scientists even suggest, as Satsuka (this volume) discusses in her chapter, that the mushrooms are best understood as 'happenings' rather than 'things'. Indeed, the Japanese word 'matsutake' is composed of two *kanji* (Chinese characters): 松 (*matsu*, pine) and 茸 (*take*, mushroom). The word not only nominates its object; it also signals the multispecies dynamics central to matsutake growth.

Our research with this fungus has taught us that attention to matsutake elusiveness can help us understand how multispecies engagements contribute to transnational worlds. In our understanding, elusiveness is not an inherent property of certain, mysterious beings. Rather, as Hathaway (this volume) explains, matsutake become 'elusive' only when someone or something wants to capture them but cannot, even while someone or something else can. In other words, elusiveness depends on forms of attraction as much as on processes of evasion or escape. A dynamic of interactions across diverse life forms, it is an effect of shifting attunements among different modes of perception and being. Notice how in the paragraph above every example of matsutake's elusiveness to humans involves other parties that desire to understand, locate, produce, consume, or otherwise engage with the mushroom. To be elusive is to be party to attraction with multiple beings across diverse sites.

Our focus on matsutake elusiveness draws attention to the complex multiparty relationships through which all beings travel, live, and grow. We ask how and why other beings become attuned to matsutake, and we examine the kinds of coordinations that emerge through these processes. We follow the charismatic pull of matsutake as they entice nematodes, insects, deer, dogs, pickers, traders, consumers, scientists, artists, and others. We consider how attraction becomes a means through which the fungus moves in and out of the rhythms of different life-ways, both human and otherwise.

Our aim is to bring questions of multispecies ontologies into dialogue with recent literature on 'global processes'. Our attention to matsutake's multispecies encounters builds on what we think of as our 'encounters' approach to globalization (Faier and Rofel 2014). We see ourselves as part of an intellectual movement exploring how cosmopolitan worlds emerge through encounters across difference (Choy 2011; Faier 2009; Faier and Rofel 2014; Hathaway 2013; Massey 1994; Rofel and Yanagisako 2019; Satsuka 2015; Tsing 2005; Zhan 2009). This approach refuses to see 'globalization' as producing either a growing global homogeneity or a set of partitioned cultural enclaves. Instead, we focus on how global processes emerge through the coming together of different ways of being and forms of attraction and desire. We have elsewhere discussed these dynamics as processes of 'world-making' (Hathaway 2013), 'friction' (Tsing 2005), 'translation' (Satsuka 2015), 'emergence' (Choy 2011), and 'cultural encounter' (Faier 2009).

In this book, we ask how we can account for the roles that more-than-human beings figure in these processes. Drawing together cases from across the planet, we illustrate that the cosmopolitan worlds of matsutake cannot be explained by the plans of any single agent or set of practices, political economic or otherwise. Instead, we explore how varied and situated multispecies coordinations knit together the diverse world-making processes through which matsutake flourish, attract, and elude. We focus on the ways that cosmopolitan matsutake ontologies emerge through the friction of everyday practical encounters, and we explore the new constellations of people, insects, trees, and other beings to which they give rise.

In the remainder of this introduction, we first introduce the research collaboration that led us to focus on matsutake elusiveness. We consider how our efforts to extend our collaboration to mushrooms and other forest beings inspired us to turn to science as a means for translating across species-being. As Kohn notes in his afterword, our collaborative research practice directly shaped our theoretical findings; the latter cannot be understood apart from the former. We then explore how our approach to global processes can offer a new perspective on multispecies ontologies by bringing matsutake into relief as a cosmopolitan mushroom multiple. We conclude by considering the research paths opened up by our wild mushroom chase.

Collaboration as a Method for Translating Across Multispecies Worlds

Matsutake came to our attention as charismatic commodities. As Faier (this volume) details, a globalized matsutake commodity chain developed to accommodate rising consumer demand during Japan's frothy economy of the 1980s. Our collaboration enabled us to track this chain across markets and forests by bringing together a wide range of area studies and scientific expertise. In 2005, we created the Matsutake Worlds Research Group (MWRG)—a team of six researchers (Timothy Choy, Lieba Faier, Michael Hathaway, Miyako Inoue, Shiho Satsuka, and Anna Tsing) with different regional foci and methodological commitments—to follow matsutake across the globe. Three of us (Faier, Inoue, and Satsuka) had encountered the mushroom during research in Japan; two in our group (Hathaway and Choy) worked in China, where the mushroom is gathered and sold for Japanese markets; one member (Tsing) had found herself at the center of a matsutake export trade in forests of the Pacific Northwest; Tsing also later went on to gather matsutake in Northern Europe; and another member (Satsuka) met local matsutake pickers in her fieldwork in the Canadian Rockies. Collectively, we had mapped a constellation of matsutake life-worlds that spanned villages in rural Japan, forests in the Pacific Northwest, science labs in Finland, trading ports in China, and high-end urban Japanese restaurants and supermarkets. We wondered what made this cosmopolitan human-fungal-arboreal landscape possible and what roles a mushroom played in linking far-flung sites.

To answer these questions, we undertook fieldwork in rural and urban settings in China, Japan, Finland, Canada, and the United States. We conducted years of participant observation and hundreds of interviews with pickers, traders, importers, scientists, government representatives, and consumers. We formed a shared archive of multi-sited research materials, some of which were gathered independently and significant segments of which were gained in joint fieldwork. We met regularly, not only to share results, but also to take pleasure in discussing new ideas as we developed and reformulated our research questions. Elaine Gan later joined the collaboration as well.

Because we depended on each other's research to make sense of our ethnographic questions, we developed ways of working together across distance and difference. We engaged in 'ethnographic echolocation' (MWRG 2009b), bouncing our observations off each other to develop multi-dimensional perspectives on shared informants, data, and research objects. We practiced textual 'poaching', what we call the process of working with and through conversations and texts produced by each other as we wrote (Choy and Zee 2015; Faier 2010; MWRG 2009a). We experimented with 'slow thinking' (Kahneman 2011) and writing 'mycorrhizally' (MWRG 2009a), dipping in and out of each other's ethnographic findings even as we developed our own perspectives. We developed new collaborative research techniques to cross geographical and disciplinary boundaries.

Our group meetings led us along different, but connected, ethnographic paths. From the outset, we decided that our collaboration would differ from the types undertaken by natural scientists, who often see themselves as working on subcomponents of a single comprehensive research plan. We did not aim to more efficiently capture component truths to place within a singular narrative about matsutake. Rather, building from our encounters methodology, we aimed to contribute a model of shared research that offered not only 'thick descriptions' (Geertz [1973] 2000) but also a collection of variegated narratives that could draw attention to similarities, differences, and uneven connections among concepts and sites.

As we explored the mushroom's multifaceted engagements, we were inspired to collaborate not only with each other but also with mushrooms and other forest dwellers. Anthropologists have long endeavored to understand life-worlds dramatically different from their own. Fungi, forests, weather, and bugs are central participants in the processes we aimed to study. We were dedicated to learning about their worlds, not just to understanding the perspectives of one human informant or another on multispecies encounters. Yet when we began our collaboration in 2005, few, if any, anthropologists had considered what such an ethnographic project might involve.[2] Anthropology (by definition, 'the study of humankind') lacked tools for accessing the dynamics of more-than-human worlds. Classical fieldwork methods were developed to collect 'facts' and 'data' from single human cultural groups. Ethnographers often saw themselves as pioneers, 'lone ethnographers' (Rosaldo 1989) working independently in their field sites. They conducted participant observation to identify objects, social structures, and cultural phenomena believed to have discernable ontologies (see Bubandt 2014 for a similar observation). If we were to learn from the mushroom to think "Other-wise" (Bhabha 1994: 91), we would need to develop a more capacious ethnographic approach. Even if mushrooms (and the other forest beings with which they grow) are important parts of human worlds, they cannot be interviewed or observed using traditional anthropocentric ethnographic techniques. We needed to be able to translate across multiple forms of being and means of expression—across modes not only of verbal or symbolic communication but also of chemical processing and attraction.

We soon realized that to gain insight into the life-worlds of non-human beings, we had to learn to think alongside scientists, who have their own strategies and technologies for accessing matsutake's sensory dexterities and communication forms. We could no longer simply treat the work of mycologists and ecologists as modernist narratives or objects of textual analysis. We needed to engage with their work in new, experimental ways. Although our collaboration used conventional interviews and fieldwork methods, we did not always restrict ourselves to them. We also spent considerable time reading scientific studies about forests and fungi, meeting with scientists, and engaging with forest beings. We learned new languages of biocommunication so that we could explore the polysemous chemical vocabularies of people, fungi, plants, and atmospheres (see Choy and Hathaway, this volume). As we did so, we learned to acknowledge the materiality of multispecies social relations beyond what our human informants claimed.

We were inspired by the matsutake scientists studied by Satsuka (this volume). These scientists develop their findings by opening themselves up to the fungus, even as they approach it as incomprehensibly different from themselves. The elusive mushroom enchants these matsutake scientists; yet, as Satsuka shows, they are not driven to try to solve its puzzle by conventional methods of objectification. Rather, these scientists cultivate an enriched sensitivity toward the mushroom by, in Choy's words, 'tending' to it. Following these scientists, we came to approach science not as a Western ideology but as a translation across world-making practices. From such a view, matsutake science became a companion account and a 'co-labor-ation', a complementary work that endeavors to translate across shared multispecies worlds.[3]

This insight also led us to approach science not as a global *lingua franca* but as a situated tool for translating more-than-human *umwelten* into languages of human knowledge and experience. Such an understanding of science enabled us to move beyond a 'two-world model' that isolates and reifies Western science and indigenous knowledge as independent cultural formations. Instead, it offers a view of science as part of geographically specific, but interconnected, zones of multispecies engagements spanning the globe.

Cosmopolitan Ontologies of the Mushroom Multiple

The chapters in this book bring both our encounters approach to global processes and our translational understanding of science into conversation with recent work on multispecies relationships. Over the past decade, scholars have looked to more-than-human worlds to challenge fixed and bounded ontologies of humanness (Haraway 2003; Hayward 2010; Kirksey and Helmreich 2010; Kohn 2013; Lorimer 2012; Ogden et al. 2013; Parreñas 2012; Tsing 2015). Inspired by post-structural critiques of fixed ontologies (e.g., Derrida 1981; Foucault 1977), these scholars have challenged binaries of 'nature' and 'culture'.

They have raised important questions about how we can understand ontology, which in a philosophical tradition from Aristotle through Derrida has focused on the distinctive 'being' of humans as opposed to the mere existence of plants and animals.

Some of our most generative insights have come from science studies scholars, such as Donna Haraway, Karen Barad, and Bruno Latour. These academics have questioned scientific assumptions that human 'being' can be understood apart from human relationships with more-than-human beings. Instead, they endeavor to account for the agency of more-than-human beings in shaping human worlds. Their work has introduced what we think of as 'onto-relational approaches'. They ask us to consider human being as produced relationally—through 'relationships' (Haraway 2003), 'intra-actions' (Barad 2007), or 'networks' (Latour 1987)—with more-than-human beings. We now recognize that 'the human is more than human' (Sagan 2011) and that "human nature is an interspecies relationship" (Tsing 2012: 141). This displacement of humans as the ultimate source of knowledge and control has allowed us to explore the roles that non-humans have played in the making of matsutake worlds. We share with these studies a commitment to treating humans and non-humans as part of joint social processes. Yet they offer little help for understanding the ways that multispecies engagements contribute to cosmopolitan social worlds.

Recently, a second challenge to traditional metaphysical approaches to ontology has emerged within the discipline of anthropology. Some refer to this move as the field's 'ontological turn' (Bessire and Bond 2014; Descola 2013; Holbraad et al. 2014; Kelly 2014; Paleček and Risjord 2013; Viveiros de Castro 1998, 2004, 2012). Developed along different lines by Eduardo Viveiros de Castro and Annemarie Mol, these approaches introduce questions of what we call 'onto-logical polymorphism' to invite a rethinking of humanness.

First, drawing on Amerindian mythology, Viveiros de Castro (1998) has developed a theory of perspectivism in which 'natural' ontologies are relational perspectives shaped by physical form. The worlds he describes are not 'multicultural', worlds in which all beings share a common nature that is regarded through different cultural points of view; rather, they are 'multinatural'. In these worlds, all beings share a common human, cultural condition that assumes different natural, corporeal states. For Viveiros de Castro, metaphysical understandings of human being that separate it from nature are limited by their ethnocentrism. He suggests that by adopting Amazonian perspectivism, we can understand being as differently grounded and thus open ourselves up to alternative ontologies.

In contrast, Mol's (2002: 32–33) approach is 'praxiographic'. She offers an understanding of being centered on everyday practice, and specifically on the practical enactments through which an object is realized. In such a view, being is a contingent and precarious accomplishment. An object does not exist in and of itself; instead, its ontology is multiple—decentered in a multitude of situated practices. Mol is interested in connections among different enactments of an object. She demonstrates that objects matter in the world precisely because of their partial connections and multiplicity. Her existential polymorphism derives from distinctive forms of social practices; for instance, different medical worlds produce different 'bodies'.

Our questions of matsutake elusiveness bring these literatures into dialogue with our work on global processes by exploring the ontologically productive engagements of multispecies worlds across transnational sites. The practices that we trace are part of worlds that center on a mushroom; these are worlds in which humans are one actor among many multispecies agents. At the same time, we focus on matsutake's elusive forms of being as geographically and historically specific. One might say that we are tracking a 'mushroom multiple', the enactments of which involve more-than-human beings. However, our approach also differs from previous studies of polymorphic ontologies. The matsutake enactments that we explore do not occur within a single bounded hospital, laboratory, network, or set of cultural practices. Rather, we track matsutake's cosmopolitan ontologies as they emerge in the overlaps and gaps of the mushroom's worldly encounters.

Our point of departure is a transnational political economic landscape linking people, places, and more-than-human beings across the globe. Our collaboration offers a multi-sited research strategy for understanding how worlds that are simultaneously local and global emerge through the multispecies engagements inspired by matsutake's chemical and sensory enticements. For example, Hathaway (this volume) builds on Jakob von Uexküll's ([1909] 2010) notion of *umwelten*, the particular sensorial capacities through which an organism experiences and engages its knowable world, by asking how different organisms' *umwelten* lead them to engage with matsutake—for instance, by attracting or repelling other beings. In contrast to Uexküll, who assumed that organisms' *umwelten* are like discrete experiential bubbles, Hathaway considers how they develop through translocal multispecies encounters.[4] Choy (this volume) offers the notion of 'attunement' to help us understand the ways in which different beings open themselves to the attractions of others. Building on what Jamie Lorimer (2007) has called 'non-human charisma' (the ways that humans become attracted to more-than-human beings), Choy argues that attunements develop within ecologies of attraction: they are arrangements of competencies that cultivate sensing bodies, enabling things and distinctions to become sensible to them. Inoue (this volume) picks up Choy's focus on matsutake's chemical charisma by showing how consumer attraction to matsutake's aroma is tied to elusive citational referents. Finally, Gan and Tsing (this volume) track the temporal coordinations through which matsutake forests grow, teasing out how translocal histories are part of the contingencies of their emergence.

Collectively, our interest lies in "interspecies intimacies" (Hustak and Myers 2012: 106) as they take shape through transcontinental links. We draw attention to the contingent worlds that emerge as different 'species' attune to, and coordinate with, each other and, thereby, enable matsutake growth. The case studies presented here—crafted in dialogue and drawing on each other's insights and research—show how matsutake worlds develop through shifting coordinations, including forms of serendipity and accident, that do not form predictable structures. We demonstrate that multispecies encounters are not defining moments of difference but provocations toward relational self-transformation as organisms respond to one another.

Chasing Wild Mushrooms: The Possibilities and Surprises of Elusiveness

Our argument builds on earlier work on matsutake worlds that our group members have published and that are now in progress. This collection is a companion piece to a larger body of work (Faier 2011; Hathaway 2014, 2015, 2016; MWRG 2009a, 2009b; Satsuka 2011; Tsing 2009, 2011, 2012, 2015; Tsing and Satsuka 2008). Our first publications considered the possibilities that collaborative research offers for multi-sited ethnographic projects. We took inspiration from matsutake to consider what it would mean to think 'mycorrhizally' (MWRG 2009a) and to explore what we could learn from a mushroom. More recently, Tsing (2015) published *The Mushroom at the End of the World*, the first of a planned three-volume set of ethnographic monographs on matsutake by members of our group.[5] Tsing's book illustrates how matsutake grows through the 'unintentional design' of forest life. Because matsutake refuses the industrial scalability central to capitalist control, capitalist modernities must work through 'salvage accumulation' to appropriate its value. Tsing shows how this mode of accumulation, a key dynamic of capitalist processes, leaves patches of ruin in its wake. But matsutake nonetheless survives—and even thrives—in such landscapes. Eluding principles of progress, civilization, and development, it offers new possibilities for life amid capitalist destruction.

This book continues these discussions by exploring emergent sites of possibility and hope introduced by matsutake elusiveness. Turning to the excitement and unpredictability of lives and worlds that escape modernist plans, we offer a collection of ethnographic strategies for chasing matsutake elusiveness and the hopeful, awkward, and mysterious worlds to which it gives rise. To chase is to take the lead from another. Chasing leads to new relationships and unforeseen paths. In these chapters, we explore how matsutake chasers—ourselves included—become subject to the objects of their pursuits. As we follow both the mushroom and its chasers, we highlight forms of attraction—commercial, aesthetic, chemical, conceptual, sensory, scientific, atmospheric, activist—that link multispecies being across the globe. Through them, matsutake come to matter to different beings in different ways.

Our matsutake chase led us beyond anthropocentric conceptions of communication and desire to account for more-than-human-centered processes of attunement and coordination that swirl around the mushroom. These processes include forms of chemical signaling among insects and plants (Hathaway, this volume), as well as atmospherics of chemical sensitivity (Choy, this volume) and experiences of sensory attraction (Inoue, this volume). They include the multispecies temporal coordinations that enable matsutake growth (Gan and Tsing, this volume) and involve the ways that such coordinations inspire the reworking of transnational political economic relations, cultural commitments (Faier, this volume), and scientific endeavors (Satsuka, this volume) in human-centered worlds. By bringing these scenes together, we offer a variegated picture of matsutake and the diverse-yet-connected worlds in which it grows. Methods of chasing are world-making projects. Chasers must learn to traverse, translate, juggle, and toggle different ontologies and modes of discernment as they follow their desired objects across worlds of practice and modes of being. In turn, chasers contribute to the emergence of new worlds through this process.

Our focus on 'the chase' (Choy, this volume) enables us to examine how an elusive being directs others' paths, even while it resists capture. It provides ethnographic traction for conceptualizing how affective, sensory, chemical, ecological, economic, biological, atmospheric, and activist relations and forms of attraction and elusiveness come to matter, intermingle, and take hold on a global scale. It is through the bundlings and bumpings of such modes of being that a planetary patchwork of multispecies engagements comes to be.

In the chapters that follow, Faier begins by introducing some of the thrills and anxieties of matsutake consumption on the Japanese archipelago. The multispecies relationships that enable matsutake growth create unpredictable harvests. Under such conditions, matsutake elusiveness—and thus its value as an elite commodity—can itself be elusive. Faier explores some of the unresolvable contradictions that inform human commitments to matsutake commodity elusiveness. She shows that in years of bountiful supply, Japanese consumers maintain their commitment to matsutake elusiveness by framing the mushroom's abundance as a 'euphoric anomaly'. Hathaway then zooms in on the role that attractions between matsutake and insects play in contributing to the mushroom's elusiveness to humans. He focuses on the mushroom's sensory engagements, such as the perception and release of chemicals, to consider how the matsutake becomes an object of attraction among multiple organisms, including insects, deer, and humans. Hathaway demonstrates how these dynamics continue to shape international matsutake commodity chains, even after matsutake are picked from the ground. At no point do the mushrooms become a commodity independent of other species and completely under human control. Hathaway shows that our grasp of them remains tied to elusive multispecies worlds, even as we consume them.

Choy then draws our attention to the ways matsutake's elusive chemistry attunes scientists and others to its life-world. Focusing on matsutake's fragrance, he argues for the 'atmospheric' as a mode of appreciation. He shows that the atmospherics of the mushroom draw scientists and others to 'tend to' its 'suspensions', its elusive states of dispersal that exist in between time and space. As we are drawn to these suspensions, the mushroom redirects our paths. Inoue follows by extending Choy's consideration of the chemistry of matsutake's elusive aromatic charisma by tracing how Japanese consumers struggle to put their olfactory experiences of the mushroom's aroma into words. She shows how the constant deferral, displacement, substitution, and inversion of efforts to linguistically represent the mushroom's fragrance sustains matsutake's associations with a similarly elusive 'Japanese-ness'.

Next, Satsuka explores how the mysteries of matsutake inspire Japanese mycologists to approach the mushrooms as *koto* (events) rather than *mono* (things). These scientists show us that matsutake can be captured only as contingent moments in which sets of multispecies relations coincide, and that they cannot be separated out as discrete objects from these relations. Finally, Gan and Tsing's piece takes our argument deep into the forest. They explore how coordinations among multiple forms of temporality enable matsutake assemblages to hold together. They focus on more-than-human scenes of entanglement in matsutake landscapes. These landscapes—sedimentations of encounters among humans and other forces and beings—are produced through forms of interspecies rhythm. Gan and Tsing present their narratives via a combination of text and diagrams that are integral to their chapter. The images visually portray processes that are generated through moments of encounter in which "rhythms of life resonate and harness each other."

Together, these chapters present a cosmopolitan 'matsutake-scape', a forum of wild creativity. In it, both human and more-than-human beings engage in diverse forms of mushroom pursuit. As they do so, their life-worlds intertwine, at once enabled and transformed by the chase. Please join us as we as pursue elusive matsutake across this patchy landscape of fragrant forests, rustic Japanese inns, bustling produce markets, swanky urban department stores, modernist government offices, ad hoc roadside stands, state-of-the-art (and not so state-of-the-art) scientific laboratories, quaint neighborhood groceries, and cozy living rooms—all of which are part of matsutake worlds.

Acknowledgments

This introduction is based on the collective work of the Matsutake Worlds Research Group. I am grateful to my collaborators—Timothy Choy, Elaine Gan, Michael Hathaway, Miyako Inoue, Shiho Satsuka, and Anna Tsing—for trusting me to serve here as a conduit for their research and ideas. My collaborators also offered extensive input and feedback on earlier drafts, as did Kathryn Chetkovich, Martin Holbraad, and Eduardo Kohn. I am deeply grateful to all for this help. I alone take responsibility for any errors and omissions.

Lieba Faier is an Associate Professor of Geography at the University of California, Los Angeles. Her first book, *Intimate Encounters: Filipina Women and the Remaking of Rural Japan* (2009), is an ethnography of cultural encounters among Filipina migrants and their Japanese families and communities in rural Nagano. She is currently writing a second book, which examines ongoing efforts to fight human trafficking in Japan. She has published in *American Ethnologist, Cultural Anthropology, Annual Review of Anthropology, Transactions of the Institute of British Geographers*, and *Environment and Planning A*.

Notes

1. A large body of literature dating back to Immanuel Kant has explored the politics and possibilities of cosmopolitanism, yet all from the perspective of human sociality (Kleingeld 2011). Such theories do not easily apply to the worlds of more-than-human beings. Instead, we use the word 'cosmopolitan' in the vernacular sense in which it refers to plants and animals that are found across the globe.
2. Our *American Ethnologist* piece (MWRG 2009a) was one of the first publications to use the term 'multispecies', which we adopted from the "Multispecies Salon" held at University of California, Santa Cruz, in conjunction with the American Anthropological Association's annual meeting in November 2006. A 2010 special issue of *Cultural Anthropology* on multispecies ethnography, edited by Eben Kirksey and Stephan Helmreich, soon followed. Over the past decade, a number of anthropologists have offered other wonderful and innovative possibilities for such an endeavor (see, e.g., Kirksey 2014).
3. For a different, but compatible, take on research as a work of co-labor, see Taguchi (2017).
4. Our interest in Uexküll joins that of several other anthropologists who have engaged with his work (Helmreich 2009; Ingold 2000; Jensen 2017; Kohn 2013).

5. Volume 2, by Hathaway, is in press with Princeton University Press, and volume 2 is being written by Satsuka.

References

Appadurai, Arjun. 1996. *Modernity at Large: Cultural Dimensions of Globalization.* Minneapolis: University of Minnesota Press.

Barad, Karen. 2007. *Meeting the Universe Halfway: Quantum Physics and the Entanglement of Matter and Meaning.* Durham, NC: Duke University Press.

Bessire, Lucas, and David Bond. 2014. "Ontological Anthropology and the Deferral of Critique." *American Ethnologist* 41 (3): 440–456.

Bhabha, Homi K. 1994. *The Location of Culture.* New York: Routledge.

Bubandt, Nils. 2014. *The Empty Seashell: Witchcraft and Doubt on an Indonesian Island.* Ithaca, NY: Cornell University Press.

Choy, Timothy. 2011. *Ecologies of Comparison: An Ethnography of Endangerment in Hong Kong.* Durham, NC: Duke University Press.

Choy, Timothy, and Jerry Zee. 2015. "Condition—Suspension." *Cultural Anthropology* 30 (2): 210–223.

Clarke, Kamari Maxine. 2004. *Mapping Yorùbá Networks: Power and Agency in the Making of Transnational Communities.* Durham, NC: Duke University Press.

Derrida, Jacques. 1981. *Dissemination.* Trans. Barbara Johnson. Chicago: University of Chicago Press.

Descola, Philippe. 2013. *Beyond Nature and Culture.* Trans. Janet Lloyd. Chicago: University of Chicago Press.

Faier, Lieba. 2009. *Intimate Encounters: Filipina Women and the Remaking of Rural Japan.* Berkeley: University of California Press.

Faier, Lieba. 2010. "Thoughts for a World of Poaching." *Cultural Anthropology.* 10 October. http://www.culanth.org/fieldsights/276-thoughts-for-a-world -of-poaching.

Faier, Lieba. 2011. "Fungi, Trees, People, Nematodes, Beetles, and Weather: Ecologies of Vulnerability and Ecologies of Negotiation in Matsutake Commodity Exchange." *Environment and Planning A: Economy and Space* 43 (5): 1079–1097.

Faier, Lieba, and Lisa Rofel. 2014. "Ethnographies of Encounter." *Annual Review of Anthropology* 43: 363–377.

Foucault, Michel. 1977. "Truth and Power." In *Power/Knowledge: Selected Interviews and Other Writings, 1972–1977,* ed. Colin Gordon; trans. Colin Gordon, Leo Marshall, John Mepham, and Kate Soper, 109–133. New York: Pantheon Books.

Geertz, Clifford. (1973) 2000. *The Interpretation of Cultures.* New York: Basic Books.

Haraway, Donna. 2003. *The Companion Species Manifesto: Dogs, People, and Significant Otherness.* Chicago: Prickly Paradigm Press.

Hathaway, Michael J. 2013. *Environmental Winds: Making the Global in Southwest China.* Berkeley: University of California Press.

Hathaway, Michael J. 2015. "Wild Commodities and Environmental Governance: Transforming Lives and Markets in China and Japan." *Conservation and Society* 12 (4): 398–407.

Hathaway, Michael J. 2016. "Rethinking the Legacies of 'Subsistence Thinking.'" In *Subsistence under Capitalism: Historical and Contemporary Perspectives*, ed. James Murton, Dean Bavington, and Carly Dokis, 234–253. Montreal and Kingston: McGill and Queens University Press.

Hayward, Eva. 2010. "Fingeryeyes: Impressions of Cup Corals." *Cultural Anthropology* 25 (4): 577–599.

Helmreich, Stefan. 2009. *Alien Ocean: Anthropological Voyages in Microbial Seas*. Berkeley: University of California Press.

Holbraad, Martin, Morten Axel Pedersen, and Eduardo Viveiros de Castro. 2014. "The Politics of Ontology: Anthropological Positions." *Cultural Anthropology*. 13 January. https://culanth.org/fieldsights/462-the-politics-of-ontology -anthropological-positions.

Hustak, Carla, and Natasha Myers. 2012. "Involutionary Momentum: Affective Ecologies and the Sciences of Plant/Insect Encounters." *differences* 23 (3): 74–118.

Inda, Jonathan Xavier, and Renato Rosaldo, eds. 2008. *The Anthropology of Globalization: A Reader*. 2nd ed. Malden, MA: Blackwell.

Ingold, Tim. 2000. *The Perception of the Environment: Essays on Livelihood, Dwelling and Skill*. London: Routledge.

Jensen, Casper. 2017. "The Umwelten of Infrastructure: A Stroll Along (and Inside) Phnom Penh's Sewage Pipes." *Zinbun* 47: 147–159.

Kearney, Michael. 1995. "The Local and the Global: The Anthropology of Globalization and Transnationalism." *Annual Review of Anthropology* 24: 547–565.

Kahneman, Daniel. 2011. *Thinking, Fast and Slow*. New York: Farrar, Straus and Giroux.

Kelly, John D. 2014. "The Ontological Turn: Where Are We?" *HAU: Journal of Ethnographic Theory* 4 (1): 357–360.

Kleingeld, Pauline. 2011. *Kant and Cosmopolitanism: The Philosophical Ideal of World Citizenship*. Cambridge: Cambridge University Press.

Kirksey, Eben, ed. 2014. *The Multispecies Salon*. Durham, NC: Duke University.

Kirksey, S. Eben, and Stefan Helmreich. 2010. "The Emergence of Multispecies Ethnography." *Cultural Anthropology* 25 (4): 545–576.

Kohn, Eduardo. 2013. *How Forests Think: Toward an Anthropology beyond the Human*. Berkeley: University of California Press.

Latour, Bruno. 1987. *Science in Action: How to Follow Scientists and Engineers through Society*. Cambridge, MA: Harvard University Press.

Lorimer, Jamie. 2007. "Nonhuman Charisma." *Environment and Planning D: Society and Space* 25 (5): 911–932.

Lorimer, Jamie. 2012. "Aesthetics for Post-human Worlds: Difference, Expertise and Ethics." *Dialogues in Human Geography* 2 (3): 284–287.

Lowe, Celia. 2010. "Viral Clouds: Becoming H5N1 in Indonesia." *Cultural Anthropology* 25 (4): 625–649.

Massey, Doreen. 1994. *Space, Place, and Gender*. Minneapolis: University of Minnesota Press.

Mitchell, Timothy. 2002. *Rule of Experts: Egypt, Techno-politics, Modernity*. Berkeley: University of California Press, 19–53.

Mol, Annemarie. 2002. *The Body Multiple: Ontology in Medical Practice*. Durham, NC: Duke University Press.

MWRG (Matsutake Worlds Research Group). 2009a. "A New Form of Collaboration in Cultural Anthropology: Matsutake Worlds." *American Ethnologist* 36 (2): 380–403.

MWRG (Matsutake Worlds Research Group). 2009b. "Strong Collaboration as a Method for Multi-sited Ethnography: On Mycorrhizal Relations." In *Multi-sited Ethnography: Theory, Praxis and Locality in Contemporary Research*, ed. Mark-Anthony Falzon, 197–214. London: Ashgate.

Nading, Alex M. 2017. "Local Biologies, Leaky Things, and the Chemical Infrastructure of Global Health." *Medical Anthropology* 36 (2): 141–156.

Ogden, Laura A., Billy Hall, and Kimiko Tanita. 2013. "Animals, Plants, People, and Things: A Review of Multispecies Ethnography." *Environment and Society: Advances in Research* 4: 5–24.

Ong, Aihwa. 1999. *Flexible Citizenship: The Cultural Logics of Transnationality*. Durham, NC: Duke University Press.

Ong, Aihwa. 2003. *Buddha Is Hiding: Refugees, Citizenship, the New America*. Berkeley: University of California Press.

Ong, Aihwa. 2006. *Neoliberalism as Exception: Mutations in Citizenship and Sovereignty*. Durham, NC: Duke University Press.

Paleček, Martin, and Mark Risjord. 2013. "Relativism and the Ontological Turn within Anthropology." *Philosophy of the Social Sciences* 43 (1): 3–23.

Parreñas, Rheana "Juno" Salazar. 2012. "Producing Affect: Transnational Volunteerism in a Malaysian Orangutan Rehabilitation Center." *American Ethnologist* 39 (4): 673–687.

Rofel, Lisa, and Sylvia J. Yanagisako. 2019. *Fabricating Transnational Capitalism: A Collaborative Ethnography of Italian-Chinese Global Fashion*. Durham, NC: Duke University Press.

Rosaldo, Renato. 1989. *Culture and Truth: The Remaking of Social Analysis*. Boston, MA: Beacon Press.

Sagan, Dorion. 2011. "The Human is More than Human: Interspecies Communities and the New "Facts of Life." *Cultural Anthropology*. 24 April. https://culanth. org/fieldsights/228-the-human-is-more-than-human-interspecies-communities-and-the-new-facts-of-life.

Satsuka, Shiho. 2011. "Eating Well with Others/Eating Others Well." *Kroeber Anthropological Society Papers* 99/100: 134–138. Special 100th anniversary issue.

Satsuka, Shiho. 2015. *Nature in Translation: Japanese Tourism Encounters the Canadian Rockies*. Durham, NC: Duke University Press.

Taguchi, Yoko. 2017. "An Interview with Marisol de la Cadena." *NatureCulture.* http://natureculture.sakura.ne.jp/an-interview-with-marisol-de-la-cadena/.

Trouillot, Michel-Ralph. 2003. *Global Transformations: Anthropology and the Modern World.* New York: Palgrave Macmillan.

Tsing, Anna L. 2005. *Friction: An Ethnography of Global Connection.* Princeton, NJ: Princeton University Press.

Tsing, Anna L. 2009. "Beyond Economic and Ecological Standardisation." *Australian Journal of Anthropology* 20 (3): 347–368.

Tsing, Anna L. 2011. "Arts of Inclusion, or, How to Love a Mushroom." *Australian Humanities Review* 50: 5–21.

Tsing, Anna L. 2012. "Unruly Edges: Mushrooms as Companion Species." *Environmental Humanities* 1: 141–154.

Tsing, Anna L. 2015. *The Mushroom at the End of the World: On the Possibility of Life in Capitalist Ruins.* Princeton, NJ: Princeton University Press.

Tsing, Anna, and Shiho Satsuka. 2008. "Diverging Understandings of Forest Management in Matsutake Science." *Economic Botany* 62 (3): 244–253.

Uexküll, Jakob von. (1909) 2010. *A Foray into the Worlds of Animals and Humans with a Theory of Meaning.* Trans. Joseph D. O'Neil. Minneapolis: University of Minnesota Press.

Viveiros de Castro, Eduardo. 1998. "Cosmological Diesis and Amerindian Perspectivism." *Journal of the Royal Anthropological Institute* 4 (3): 469–488.

Viveiros de Castro, Eduardo. 2004. "Exchanging Perspectives: The Transformation of Objects into Subjects in Amerindian Cosmologies." *Common Knowledge* 10 (3): 463–484.

Viveiros de Castro, Eduardo. 2012. "Cosmological Perspectivism in Amazonia and Elsewhere." *Masterclass Series 1.* Manchester: HAU Network of Ethnographic Theory.

Wang, Yun, Ian R. Hall, and Lynley A. Evans. 1997. "Ectomycorrhizal Fungi with Edible Fruiting Bodies: *Tricholoma matsutake* and Related Fungi." *Economic Botany* 51 (3): 311–327.

Zhan, Mei. 2009. *Other-Worldly: Making Chinese Medicine through Transnational Frames.* Durham, NC: Duke University Press.

Chapter 1

EUPHORIC ANOMALY
Matsutake's Elusive Elusiveness in 2010 Japan

Lieba Faier

This book builds from the simple observation that elusiveness can reveal as much as it eludes. Yet because, by definition, elusiveness points to that which cannot be understood or attained, it also poses an ethnographic challenge: how can we track that which endeavors to escape? This first chapter takes up this question by exploring some of the unresolvable contradictions that inform human commitments to matsutake commodity elusiveness. Focusing on these contradictions helps us track matsutake elusiveness because it draws attention to a key (but also inconsistent and incongruent) way that humans account for the uncontrollable dynamics of multispecies relationships. As illustrated in subsequent chapters by Matsutake Worlds Research Group (MWRG) members Michael Hathaway, Elaine Gan, and Anna Tsing, matsutake growth depends on forms of attraction and coordination between human and non-human beings,

Notes for this chapter begin on page 33.

and thus on processes that are not fully understood by humans and that lie outside human control. Building on my collaborators' findings, this chapter argues that to maintain their commitment to matsutake elusiveness, Japanese consumers must account for the unpredictability of matsutake availability in self-contradictory ways.

Consider what happened in 2010. When I arrived in Japan in October of that year—the middle of the domestic matsutake season—I found matsutake buyers and sellers waxing euphoric over an unexpectedly large domestic harvest. Japanese news and popular media widely discussed the *daihōsaku* (bumper harvest). People whom I interviewed reported that prices of matsutake from Nagano and Iwate prefectures (the two largest suppliers) were 30 percent less than, if not half of, recent years, creating a buying frenzy (see also *Asahi Shimbun* 2010a, 2010b; Nippon Terebi 2010).[1] One local news program during my stay explained that matsutake customers, pickers, and shop owners alike were "shrieking" with delight at the abundant supply and low prices of the mushrooms (Nippon Terebi 2010). Matsutake shops in Nagano were crowded with buyers. Restaurants serving fancy matsutake course menus were full of customers. The commentators on the news program repeatedly used the word *zeitaku* (extravagant, luxurious) when discussing both the traditional and the unusual ways people were eating domestic matsutake that year: grilled with *sudachi*, a Japanese citrus; in clay-pot broth; in sukiyaki; tempura-fried; soaked in sake; and even in a Japanese-style matsutake pasta (ibid.). How and why does a commodity that is valued in large part for its perceived elusiveness maintain that value at a time when people are celebrating its relative abundance and accessibility?

Some might argue that the euphoria surrounding the 2010 bumper harvest was a natural response to an anomalously large supply of an ordinarily elusive commodity. Indeed, as discussed below, in 2010 this idea circulated widely among Japanese consumers and in popular media. However, to assume as much would be to naturalize precisely what needs to be explained: the domestic matsutake harvest could be considered anomalously abundant only if the mushrooms had been considered elusive in the first place.

In this chapter, I explore how domestic matsutake have come to be viewed as elusive commodities in Japan since the end of World War II. I consider what this perception both reveals and hides. I suggest that understanding commitments to domestic matsutake elusiveness requires attention to historical, cultural, and geographical processes through which the mushrooms have been 'specified'—at once identified and produced (Choy 2011)—as unique and inaccessible. As Hathaway and Gan and Tsing illustrate in later chapters, these processes of specification are natural-cultural processes. They depend on the relationships among trees, fungi, insects, weather, humans, and other beings. In this chapter, I historicize how transformations of natural-cultural landscapes have inspired the notion that matsutake is an elusive commodity, and I consider how and why Japanese media outlets and consumers maintained the mushrooms as such.

Matsutake are not universally appealing. David Arora (1986: 191), an American mycologist and author of popular guides to wild mushrooms, has famously described the smell of matsutake as a "provocative compromise between 'red hots' and dirty socks." Even in Japan the mushrooms are often described as an 'old-fashioned' or 'acquired' taste. Matsutake mushrooms' appeal as a gourmet food is linked to their status as 'wild'. As discussed in the introduction, matsutake have never been cultivated, and thus their growth is viewed as beyond human understanding and control. Over the past 70 years, prices of domestic matsutake have risen due to declining matsutake yields on the Japanese archipelago, reinforcing notions of their elusiveness. However, surprisingly large harvests and drops in price occasionally happen, as was the case in 2010. Moreover, in response to growing consumer demand, large amounts of matsutake began to be imported to Japan in the 1980s. By 1992, 90 percent of matsutake purchased in Japan were imported (Sterngold 1992). Given both the existence of large imported supplies and the potential for an abundant harvest, how have domestic matsutake maintained their commodity value as elusive?

We know that value is not inherent to commodities. It is produced through social relationships, including those of production and consumption (Appadurai 1986; Caldwell 2002, 2004; Callon et al. 2002; Chin 2001; Foster 2007; Mankekar 2015; Patico 2002; Tsing 2013; Yanagisako 2002). In recent years, scholars have helped us understand how the consumer appeal of commodities rests in part on their ability to cultivate intimate, affective responses in consumers and to thereby incite desire (Foster 2007; Mankekar 2015). Some have examined how governments, corporations, and marketing specialists try, with varying success, to emotionally attach consumers to their products and thereby maintain their exchange value through associations with national, class, and other identities (Creighton 1997; DeSoucey 2010; Foster 2007; Shankar 2012). Other scholars have stressed the role of consumer agency in the production of commodity value, including how national, racial, and cultural associations make products appealing to consumers (Caldwell 2002; Callon et al. 2002; Condry 2006; Foster 2005, 2007; Mankekar 2015). Still others have considered the key role that food plays in how people craft senses of national 'selves' and, correspondingly, 'Others' (Appadurai 1988; Boisard 2003; Leitch 2003; Ohnuki-Tierney 1993).

As discussed below, discourses of nationalism and class play important roles in the articulation of matsutake as an elusive commodity. However, earlier studies have tended to focus on the ways that nationalist, diasporic, and class identities are produced through access to commodities or through consumption of them. Less attention has been paid to how these figure into the cultural practices through which commodities assume affective value as elusive, inaccessible, and elite. Indeed, anthropological discussions of the elusiveness of objects—for instance, in regard to their 'aura' (Benjamin 1968), 'distinction' (Bourdieu 1984), and 'symbolic densities' (Weiner 1992)—have centered on their exclusion from, or inaccessibility to, commodity relations. The value of elusive commodities tends to simply be attributed to economistic assumptions about supply and demand. Consequently, less is known about how commodities come to acquire value on account of their attributed elusiveness and the practices through which this association is maintained.

One recent set of studies has highlighted the ways that artisan food producers either capitalize on, or distance themselves from, nationalist associations to cultivate the elite distinction of their products. These studies show that the identification of cheese, chocolates, and 'slow' foods as 'authentic' rests on the 'craft' involved—the methods and materials through which they are made and that may challenge dominant discourses by playing on nostalgia for an aestheticized, pre-industrial work ethic or an alternative set of values (Leitch 2003; Paxson 2008; Terrio 1996). Some studies of artisan production have also considered how the distinction of specialty food products like wine can be tied to geographical features, such as *terroir* (Trubek 2008). However, the value of a 'wild' product like matsutake cannot be attributed to producers' special techniques or their selection of a specific region of production. Nor can it be based on corporate branding practices. Rather, it must be tied to the multispecies coordinations through which the product grows—elusive processes that, I argue, can also and unexpectedly betray its status as inaccessible.

In this chapter, I explore how these interactions played out during the matsutake bumper harvest of 2010. I consider both the ways in which matsutake have come to be valued as elusive and the contradictions that have enabled its status as such. In this regard, one might say that this chapter explores the elusiveness of matsutake commodity elusiveness. I show that, on the one hand, domestic matsutake have become desirable commodities for Japanese consumers because they have been identified as inaccessible due to their dependence on multispecies relationships. On the other hand, the multispecies relations through which matsutake grow do not always support the mushroom's fetishization as inaccessible.

I suggest that, in 2010, media outlets and consumers endeavored to resolve the contradictions of Japanese matsutake commodity elusiveness by presenting their widespread availability as a euphoric anomaly. Media outlets and consumers did this by situating the mushrooms within specific, bounded temporal and spatial frames: the annual local season, the developmental time of the nation-state, and the individual human lifetime. These framings produced a contradictory ontology of the mushrooms as elusive precisely because their status as such depended on the participation of beings and forces outside human control.

My argument is based on participant observation and interviews with pickers, traders, consumers, and retailers that I conducted in rural Nagano and Tokyo in 2005, 2006, 2007, and 2010. Interviews in 2005 and 2006 were undertaken with my MWRG collaborator, Anna Tsing; those undertaken in 2007 and 2010 were conducted independently. I also include research based on studies of Japanese popular media, including newspapers, magazines, and television programs. Finally, I draw on conversations with, and the research and writing of, my MWRG collaborators since 2005.

In the remainder of this chapter, I first consider how matsutake came to be perceived as an elite and expensive commodity in the decades following World War II. I then examine the temporal and spatial scales through which the mushrooms' elusive status was maintained despite the unexpectedly large harvest in 2010. I suggest that by relying on spatial and temporal scales to frame the 2010 matsutake harvest in Japan as an anomaly, consumers were able to at once enjoy the relative accessibility of the mushrooms and maintain domestic matsutake's elite status as elusive commodities.

How Common Matsutake Garnered Charisma and Became Elite

The commodity value of domestic matsutake as elusive depends on perceptions of the mushroom's increasing inaccessibility to average consumers, whose desire for the mushroom exceeds supplies available to them. These perceptions of inaccessibility developed in the context of a specific history of shifting human-forest relations, which occurred alongside discussions of the mushroom as an elite commodity in popular media in the years following World War II.

During the seventeenth century, matsutake became a widely available and consumed food in mountain communities on the Japanese archipelago. The mushrooms were hardly considered an elite or elusive product. Rather, political economic changes and accompanying shifts in forest use created new human-forest relations, which resulted in a dramatic increase in matsutake growth and, correspondingly, consumption (Arioka 1997). Matsutake, like its common host, the red pine, grow best in rocky, nutrient-poor soil. In the early seventeenth century, the Tokugawa government unified domains on the islands of Honshu, Kyushu, and Shikoku into a loose confederation of states. Old-growth forests were felled for urban development and monumental construction projects, and red pine forests took their place, offering new hosts for matsutake fungal expansion. Communities in mountainous areas used nearby forests for fuel, fertilizer, and construction materials. As they gathered leaves and grasses, they encouraged matsutake growth, and the numbers of mushrooms in the mountains increased dramatically (Arioka 1997; Yamanaka 2011). By the mid-eighteenth century, these shifts in forestry practices had made matsutake a food source in these areas that was considered too common to be served in restaurants (Arioka 1997).

As rural communities continued their relationships with surrounding forests through the nineteenth and into the twentieth centuries, the numbers of matsutake increased (Arioka 1997). Poor residents of mountain communities relied in part on wild vegetables and mushrooms for their food, and sales of the mushrooms (and their picking rights) provided an important form of income that offset taxes and contributed to the construction of local schools (ibid.). After the establishment of a Japanese nation-state in the late nineteenth century, transportation developments enabled wider access to the mushrooms with the result that they could be enjoyed throughout the country. By the mid-1920s and through the 1930s, mainstream Japanese women's magazines published matsutake recipes in their fall issues nearly every year, suggesting that the mushrooms were regularly prepared for meals at home (see, e.g., *Fujokai* 1936; Ichige 1930; Ichinoe 1925; Inada 1939; Kamei 1938; Ono 1925). The harvest in 1945 is said to have been so large that the mountains seemed covered in white buds. Imperfect matsutake were simply discarded, and caps that were already open were left in the forests to rot (Arioka 1997).

In the years following World War II, matsutake supplies on the Japanese archipelago—and the status of the mushroom—began to shift. Political economic changes affected multispecies relationships that had previously enabled matsutake to flourish. The decline of agricultural communities, the replacement of wood and other biomass fuels with oil and gas, and the importation from overseas of cheaper construction materials—all led to the neglect of Japanese pine forests. As a result, the soil in the forests grew too rich for matsutake to thrive (Satsuka 2014). Also shaping these multispecies relations were histories of trade and the reluctance to treat affected trees on military bases. Global warming trends also played a role by introducing and encouraging the spread of nematodes that kill red pine, matsutake's primary host (Faier 2011). Pine trees in Japanese forests became increasingly vulnerable to pine wilt disease on account of both rising temperatures and the decline of matsutake, with which they grow symbiotically. Matsutake forests became overgrown, abandoned, or decimated. Industrial developers targeted them for golf courses, housing developments, factories, and industrial waste sites (Saito and Mitsumata 2008; Satsuka 2014). By the mid-1950s, harvests had begun to decline (Arioka 1997).

As matsutake harvests continued to decline on the Japanese archipelago over the ensuing decades, the mushrooms increasingly garnered charisma as an elusive and thus elite food. This status was cultivated by popular news media, which focused on the rising prices and increasing inaccessibility of matsutake for the average consumer. Beginning in the 1950s, for instance, national newspapers began to explicitly comment on the inaccessibility of matsutake to 'common people' (*shomin*). An article in *Yomiuri Shimbun* (1953) explains that, despite the year's large yield, prices are staying high, adding that "it doesn't look like common people will be eating them."[2] Such comments recurred throughout the following decades, especially in years of low harvest. For instance, in the late 1970s, in a year of a poor harvest, the *Asahi Shimbun* (1978) printed observations from matsutake shops that reinforced ties between the mushrooms and elite corporations: "Only corporate agents are buying them to give as gifts. Housewives just pass on by. This year, it does not look like matsutake, which year by year are getting farther and farther away from common people's mouths, will be on the tables of ordinary households." Even in 2000—a time of economic recession when prices of the first shipments were lower than they might have otherwise been, given the limited and late appearance of the mushrooms—a national newspaper reported: "As usual, it looks like for common people's mouths they will be a distant taste of autumn" (*Yomiuri Shimbun* 2000).

These newspaper articles are part of an annual routine in Japan that includes media reporting on seasonal processes, such as cherry blossoms, across the archipelago. Through such practices, the Japanese popular media works as a regular cultural broker for matsutake commodity elusiveness, cultivating the mushroom's charisma as desirable and elite, yet inaccessible to most. As popular media circulated comments about the inaccessibility of matsutake to common people, it also cultivated associations of the mushroom with wealthy and high-status groups and practices. Consuming the mushrooms came to be desirable and something that promised membership in an elite class. In the post-war years, new corporate practices created new class identities in relation to consumption. A culture of connoisseurship developed around matsutake. The most perfectly formed and expensive matsutake came to be used by corporations as high-status gifts or were served at *ryōtei*, exclusive restaurants where power elites dine in private rooms to conduct negotiations and cement alliances (Sterngold 1992). Others with means consumed small amounts of lesser-quality mushrooms when their budgets allowed.

The elite status of the mushrooms was also cultivated by national marketing campaigns. For example, fall advertisements for upscale sake featured the mushrooms as a suitable accompaniment—in contrast, for example, to beer ads that featured the working-class fall food of grilled *sanma* (mackerel pike) (Andoh 1995). The advertisements and statements in national newspapers also encouraged feelings of longing for the mushrooms, fostering their affective pull. For example, when explaining about the year's small harvest and high prices, a 1978 newspaper article stated: "It appears that matsutake season will end with common people just 'gazing longingly'" (*Asahi Shimbun* 1978).

If matsutake's status as an elite commodity developed alongside declining yields in Japan following World War II, it also reflected the contradictions of efforts to 'modernize' Japan during this time. The 1950s to 1970s were known as decades of 'high growth' and 'income doubling' in Japan. National identity was premised on the idea that modernization had enabled all Japanese citizens to become middle class. After the deprivations of the war and its aftermath, Japanese citizens' newfound ability to purchase commodities, and particularly high-status ones, became a key symbol of a shared national success (Kelly 1986). However, at the same time that rising incomes fostered increased demand for status items, Japan's economic growth in these decades produced wealth inequality. Out of these economic changes emerged an elite class that could afford to consume expensive items and a larger group of others that could not. Laments of meager matsutake harvests and high prices in the news reports during these decades reflect this contradiction, which underpins notions both of a middle-class Japan and of matsutake commodity elusiveness to 'common' people.

Geographies of National Mushroom Development

Against the backdrop of this post-war history, the lower prices of domestically harvested matsutake in 2010 Japan offered 'common' consumers an unexpected and exciting opportunity to access a food marketed as elite. The appeal of the mushroom also reflected aspirations and desires that went beyond an individual's class identity to an affective discourse of Japaneseness as a prestigious, desirable identity on a global stage. In this section, I turn to the paradoxical way that matsutake commodity elusiveness links high culture and modernity in Japan with nostalgia for Japan's vanishing rural past, a discourse that also self-consciously emerged in Japan after World War II (Ivy 1995). This discourse metonymizes the rural as the national just as it contradictorily maintains a distinction between the two.

During the 1980s (the height of Japan's 'bubble economy'), contradictions between matsutake's elite associations as a commodity and lamentations of its inaccessibility to common people were tempered through the creation of hierarchical classes of the mushrooms. Matsutake from some regions (Tanba, Hiroshima) were considered more desirable than those from others (Nagano, Iwate). Initially, elites consumed the more prestigiously sourced ones, while lesser matsutake were left to those with fewer means. Then, as incomes increased in Japan and matsutake demand outpaced domestic supplies, large numbers of the mushrooms began to be imported into the country. These sometimes abundantly available matsutake were priced considerably below those found on the Japanese archipelago. Stores and restaurants were expected to label the sources of matsutake; if the mushrooms were not explicitly labeled, storeowners were to provide this information upon request. Like many other imported products, such as rice and cars, imported matsutake came to be viewed as a lower class of the mushrooms, one that was inferior to, and less desirable than, all domestic ones (Ohnuki-Tierney 1993). Perceptions of the elusiveness of Japanese matsutake in 2010 Japan depended on this expanded hierarchy in which domestic matsutake were believed to be qualitatively different from imported ones.

In fact, matsutake from North America are of a variety that is different from those harvested in Japan. Whether or not they are more or less delicious is a matter of preference. For example, one middle-aged Japanese-American woman whom I interviewed in Japan fiercely argued that the matsutake that she had grown up picking in the Pacific Northwest were far better than the Japanese ones that she had eaten while living in Tokyo. However, the variety of matsutake found in China and Korea is identical to that found in Japan. On what grounds could distinctions between Japanese and, relatively abundant, Korean or Chinese matsutake be justified?

The privileging of matsutake harvested in Japan over those from China, Korea, and elsewhere was at one time argued on the grounds of mushroom quality, including the temporalities of mushroom freshness, shriveling, and decay. Matsutake are considered best when they are eaten soon after picking—optimally within 24 hours, before the fragrance fades and the mushrooms begin to dry out. Highly efficient postal and delivery services enable matsutake to be shipped within Japan overnight (if not sooner). Foreign matsutake must travel farther distances to reach consumers and could take longer. In this regard, the privileged value attributed to domestic matsutake was at one time linked to customary ways of eating and appreciating the mushrooms related to matsutake life cycles and shelf life. However, with developments in airfreight technologies over the past couple decades, imports from Korea and China can now take similar amounts of time to arrive in Japan when compared to domestic mushrooms traded across the Japanese archipelago. A number of consumers and sellers told me that the differences between Japanese and Korean and Chinese mushrooms were negligible, if they existed at all. In Nagano, now viewed as a prime source of highly valued domestic matsutake, I heard stories about imported mushrooms being pawned off as domestic ones at mountain roadside stands. The privileged status of domestic matsutake thus cannot be attributed simply to the quality of imported mushrooms. Rather, it must be understood in relation to the affective pull and cultural capital that domestic products command in Japan through their association with the contradictions of nationalist discourses of Japanese culture and landscapes and, correspondingly, pure and essential forms of Japaneseness.

The privileged affective value accorded to domestic, as opposed to foreign, matsutake stems in part from the ways the mushrooms have been incorporated into developmental narratives of Japanese national culture and landscapes over the past several decades. Nationalist associations are not new for the mushrooms. For example, in the 1930s, when matsutake were still widespread in the mountains, the Daiki-Sankyu railroad company aimed to attract Japanese customers for its new routes in the Nara area with pamphlets that identified the stations close to matsutake mountains: "The fall is for going to matsutake mountains ... Let's train in the mountains! A campaign to train minds and bodies of the nation!" (Murakoso and Tanesaka 2013: 358–359).[3] The 1930s were the height of Japan's imperial projects in Asia, and this military nationalism was echoed in women's magazines. For example, the October 1938 issue of *Fujokai* (Women's World) inserted recipes for 'ten flavorful matsutake dishes' (Kamei 1938) between stories about courtships with soldiers in combat and articles related to the ongoing war in (and the occupation of) China. Presumably, these recipes could be used to please one's fiancé or husband when he returned home. During this time, the large matsutake harvest was presented as a collective opportunity to contribute to the military effort, and matsutake were viewed as a special and appealing seasonal product (like *sanma* in the aforementioned ad). However, despite the nationalist associations, the mushrooms were not necessarily viewed as elite in the way they later came to be. Indeed, it was their availability to a general population that informed their nationalist appeal.

During the post-war years—as matsutake yields declined on the Japanese archipelago in the face of the interconnected social, ecological, and political economic shifts—new, self-contradictory discourses of urbanization, modernization, and class elitism introduced novel ways of incorporating the mushrooms into narratives of nation-ness. These discourses linked the mushrooms with a traditional, rural, and essentially Japanese way of life that was vanishing in the face of Japanese modernity. As explained above, matsutake forests developed through relationships with rural communities, which encouraged the growth of the mushrooms as villagers coppiced and cleared the forest ground while gathering wood and leaves for fuel and fertilizer (Satsuka 2014). These forest-village relationships are often described as 'traditional' satoyama relationships (Saito and Mitsumata 2008). The word 'satoyama' is a compound comprised of the characters for village (里, *sato*) and mountain (山, *yama*). It refers to both the woodlands that surround a mountain village and a settlement pattern that involves living in upland valleys while cultivating the lower slopes of the mountains. For urban consumers, domestic matsutake consumption today references this way of life: it evokes nostalgia for a lost Japanese tradition in the face of modernity and for the communal and collective identities associated with it.

In step with these discourses of vanishing Japanese traditions, the declining number of matsutake found in Japanese forests has been the subject of nostalgia related to the losses sustained on account of Japanese modernity and thus Japanese nationalism. Books and articles about matsutake published in the 1990s and early 2000s sometimes cite the mention of matsutake in the Kojiki and Manyōshū, classical texts often invoked in nationalist discussions of the ancient origins of Japanese culture. These texts mention the use of the mushrooms as part of a ritual to deities or their circulation as a gift among aristocrats and lords. When invoked, these discourses articulate nationalist narratives of a timeless, elite Japanese tradition, as well as histories of land use in Japan.

Marilyn Ivy (1995) has argued that nationalist discourses of Japanese modernity rest on contradictory notions of a 'vanishing' traditional Japanese past. Writing in the early 1990s, Ivy focuses on how Japan's national economic successes produced unease about the stability and transmission of national culture. She explains that these anxieties were palpable in the fetishized discourses of vanishing Japanese traditions that produce at once, paradoxically, a nostalgic longing and a sense of disavowal. The charisma of matsutake elusiveness rests on this contradictory longing for the losses associated with Japanese modernity.

One manifestation of this longing has been matsutake tourism, which fits within a more general romanticization of Japan's agrarian heritage and a traditional and essentially Japanese rural way of life for those living in more urban areas (Creighton 1997: 241; see also Ivy 1995; Kelly 1986). Many hotels in matsutake regions offer matsutake course menus and matsutake picking to experience the fall beauty of Japanese mountains. Other areas more explicitly celebrate national matsutake histories. For instance, since 1990, the city of Oita in Gunma prefecture has held a fall procession to share the history of the "Matsutake Journey" (*matsutake dōchū*). Residents dress in Edo period costumes and carry boxes of matsutake and a matsutake palanquin (*kago*) across the city to celebrate Kanayama mountain, where matsutake were at one time plentiful, and to re-enact how, beginning in 1629, the local lord would bring the mushrooms as tribute to the shogun (*Yomiuri Shimbun* 2009).

Others have privileged domestic matsutake in relation to the heritage and future of the nation. Such concerns have come to the fore in recent years on account of the liberalization of the Japanese economy and corresponding job losses. One matsutake scientist in Nagano explained to me the importance of the domestic matsutake cultivation projects in which he is involved: "The Japanese market should be supplied with matsutake produced in Japan … Food should be produced and grown where it is eaten. For example, Japan produces and sells cars, and in exchange we buy food from the United States and China. But if car production stops, we'll soon run out of food … It is in the country's best interest to increase production levels of matsutake within Japan."

These discussions of matsutake commodity elusiveness tie in with nationalist hierarchies of development and progress, which widely circulate in Japan and position the country as superior to those places perceived as being 'less developed', such as China and Korea (Faier 2009). As discussed above, these hierarchies position Japanese products as more desirable than those from other countries to justify their higher prices. In 2005, a matsutake importer captured this sentiment in an interview with Anna Tsing and myself, explaining that the "'made in Japan' type of idea is very important. For instance, the same vegetable crop, if it's not domestic but [even of] superior quality … can [still] only get half [the] price. You know, it's not just related to the value of the produce." Such comments echo earlier published sentiments. For instance, in a 1992 newspaper article, wholesale dealers at Tsukiji market explained that although they felt that the imported mushrooms were just as good, "real purists want just Japanese mushrooms because of the feeling they get" (cited in Sterngold 1992). This 'feeling' is tied to a sense of tradition and the notion that matsutake are "ineffably Japanese" (ibid.).

Yet at the same time that domestic matsutake have been valued for their nationalist associations and rural ties, it has been social and political economic changes in rural areas since the end of World War II that have altered the human-forest relations that facilitate matsutake growth. In other words, the very processes that enabled modern nationalism in Japan have altered the human-nonhuman relationships that support matsutake. In the next section, I turn to the ways that consumers and media outlets in 2010 resolved these contradictions of classism and nationalism underlying Japanese matsutake value by spatially and temporally situating the harvest within local and national seasons and characterizing it as an anomaly—a 'once in a lifetime' occurrence.

Anomalous Ways for Resolving Elusive Contradictions

In 2010, contradictions underpinning the elusiveness of domestic matsutake were forced into relief. As I have explained, the value of these matsutake rests on notions that the mushrooms are inaccessible to most consumers and therefore elite. This valuation also supports notions of class privilege and is underpinned by Japanese national modernity and assertions of superiority toward other parts of Asia. However, on account of the 2010 bumper harvest, not only was there a relative abundance of domestic matsutake, but also the mushrooms were reported to be in some cases cheaper than Chinese and Korean ones. Some consumers explained that they were disoriented by finding domestic matsutake priced below imported ones: "It feels kinda strange," reported one woman (Nippon Terebi 2010). However, most consumers chose to look past the contradictions of the situation. They declared the year to be an anomaly in order to maintain a sense of Japanese matsutake elusiveness. To do this, consumers and media outlets situated the harvest in particular temporal and spatial frames. These frames enabled consumers to enjoy the pleasures of ready access to what was believed to be an elite and high-status commodity while maintaining the distinction of Japanese matsutake from others—even if the former were relatively cheap and abundant.

First, consumers and media outlets situated the mushrooms in a local matsutake season, removing them from the spatio-temporalities of the transnational production chains that today enable matsutake consumption. Even in the context of a dwindling supply of matsutake on the Japanese archipelago, the bumper harvest of 2010 was not entirely unprecedented. On occasion in the past decades, bumper harvests of Japanese matsutake had caused prices to plummet. Moreover, beyond these isolated instances, perceptions of matsutake elusiveness are puzzling in the context of vast global supplies. Theodore Bestor (2001) uses the term 'timescape' to characterize the complex temporal structure through which production processes and markets are coordinated across the globe to accommodate year-round demand for a given product. Timescapes are production strategies for overcoming the limitations of local seasons. They enable constant (i.e., seasonless) production on a geographically linked (i.e., global) scale. The global search for a matsutake supply has created a 'matsutake timescape', offering a flexible and sometimes copious supply of the mushrooms that are imported into Japan during much of the year.

Notably, a large supply of imported matsutake, which is sometimes available, is not necessarily considered remarkable or even desirable. For instance, in 2005, Anna Tsing and I interviewed a matsutake importer in the Kantō area who described the financial losses he had incurred one year on account of huge and unexpected harvests of Korean matsutake, which flooded the market. He also spoke of the matsutake fatigue some consumers now experience because significant quantities of relatively inexpensive Chinese mushrooms have become available in Japan months before the Japanese matsutake season starts.[4] In the face of this abundance, consumers had to remove the 2010 domestic harvest from this global timescape and situate it in a local season to maintain the mushroom's status as elite and elusive. As discussed above, one way that sellers did this was by distinguishing between Japanese and 'foreign' mushrooms.

However, bracketing global timescapes was not just a matter of differentiating the matsutake supply on parts of the Japanese archipelago from those found elsewhere. It also involved metonymizing supplies from particular regions in Japan as 'national supplies'. On account of geographically variable conditions within Japan, pine wilt disease has affected different regions in different ways. Mountainous areas most famous for matsutake, such as Kyoto or Hiroshima, were affected early on and have been particularly hard hit. Generally, trees in colder regions of Japan, such as Iwate and Nagano, were affected later, that is, in the late 1970s and early 1980s, and have suffered less damage (Futai 2008). Even within these regions some areas have been harder hit than others due to variations in temperature. Generally colder temperatures prevent or limit the spread of the disease by killing off the vectors. In addition, some regions, such as Iwate, have to some degree succeeded in preventing its spread (Kamata 2008). Thus, portions of Nagano and Iwate have in recent years supplanted Kyoto and Hiroshima as dominant producers of the mushrooms, while the small amounts of mushrooms found in Kyoto and Hiroshima sell for exorbitantly high prices.

As suggested above, in the past, regions of Japan competed over matsutake harvests and were ranked according to the quality and quantity of their mushrooms. However, as matsutake have become unavailable in regions that had been famous for them, perceptions of 'local' availability have expanded to include the entire archipelago. As was the case in 2010, harvests from select regions of Nagano and Iwate have come to represent, and are extrapolated to stand in for, a circumscribed domestic/national supply. As a matsutake importer explained to me: "When you call something seasonal, you are actually referring to your *local* supply. In Japan … it starts from the north, such as Hokkaido or Iwate prefecture, which is normally … mid-September to late September." The importer overlooked the historically famous but now low-producing matsutake regions of Kyoto and Hiroshima, instead conflating matsutake availability in portions of a few prefectures in Japan (Hokkaido, Iwate, and, by implication, Nagano) as a single 'national' qua local supply. As pickers, importers, consumers, and popular media metonymized particular local harvests as a domestic/national supply, they articulated this spatial scale and the notion of an anomalous Japanese harvest as a 'local' seasonal product.

The euphoria surrounding the domestic harvest in 2010 was also premised on removing the harvest from the temporal rhythms of forests and mountains and identifying the yield as a once in a *human* lifetime experience. As Elaine Gan and Anna Tsing (this volume) demonstrate, although fungal lives (including a fungus's fruiting cycles) share a sense of seasonal cyclicality with human lifetimes, fungi also have their own rhythms and schedules. Consumers (and scientists) have yet to discern whether, from the perspective of the fungus or other forest creatures, a harvest is abundant or scarce. One way that media outlets and consumers contained this uncertainty was by focusing on matsutake harvests within short-term human time frames as opposed to longer fungal histories. In 2010, the supply was compared only to similarly local harvests from previous years. For instance, pickers focused on recent matsutake seasons when they had struggled to find even one mushroom, whereas in 2010, clusters of 10, 20, sometimes 30 large, well-formed mushrooms caps were being found by even inexperienced pickers, including elementary schoolchildren (Nippon Terebi 2010; Suzuki 2010). News reports looked to the previous year when local elementary schools in Japan had canceled their annual matsutake-picking activities on account of a poor harvest (Suzuki 2010). In comparison, they marveled at the large quantities of matsutake that elementary students were finding in October 2010 and showed tables of the mushrooms laid out in front of the children, explaining that the mushrooms would be served that year in school lunches (Nippon Terebi 2010).

Consumers also discussed the significance of the event in the context of their individual life spans: "It's been 40 years since I ate domestic, local matsutake," one man enthused (Nippon Terebi 2010). Another interviewee, a middle-aged man enjoying an elaborate gourmet menu of matsutake, said: "I am truly happy today. It is the first time in my life that I have finally fulfilled my dearest wish of eating matsutake dishes to my heart's content" (ibid.). This euphoric once in a lifetime experience of matsutake abundance rested on the disjuncture between matsutake's 'fungal clock' (Gan and Tsing, this volume) and a human lifetime (around 83 years in 2010 Japan). However, consumers could not ignore the fact that this disjuncture could produce not only scarcity but also abundance. Matsutake's commodity value lay in the multispecies elusiveness that enabled both to occur.

Conclusion

In this chapter, I have considered how the very multispecies relationships that have enabled matsutake to be valued as an elusive commodity can result in its widespread abundance. In bountiful times, such as in 2010, these contradictions have to be resolved to maintain commitments to matsutake elusiveness. In 2010, by framing the harvest as a euphoric anomaly within certain spatial and temporal frames, Japanese consumers reinforced notions of matsutake inaccessibility while enjoying its abundance.

Philip Deloria (2004) argues that there are different ways for framing unanticipated events. He contrasts the anomalous with the simply unexpected. The latter, he explains, resists categorization and throws expectations into question. The former does the opposite. By defining the unnatural and odd, it reinforces expectations, recreating and empowering dominant categories. In this way, the anomalous reinforces expectations by accounting for that which does not fit them. It allows for the co-existence of contradictions by enabling the exception to hold the rule.

Japanese consumers and news media framed the 2010 domestic matsutake harvest as an anomalous event by temporally and spatially situating it within frameworks of the individual human lifetime and the local landscape metonymized as a national landscape. In the context of these temporal and spatial scales, the class and nationalism-based contradictions that also underpin matsutake commodity elusiveness could be ignored. Thus, the unanticipated large supply of Japanese mushrooms could be euphorically celebrated as an aberration that would maintain the mushroom's elite status.

In this example, we see how the affective draw of domestic matsutake elusiveness rests on investments in a national landscape, including the spatiotemporal frames that underpin it. However, we see that it also depends on investments in the unpredictability of matsutake supply on account of multispecies relations, which are perceived as irreparably harmed in the face of modernity. As explained above, matsutake were incorporated into narratives of nation-ness in the post-war period as part of emerging discourses of urbanization and modernization, viewed as responsible for a vanishing traditional rural way of life. These narratives embed significant contradictions that both reinforce and undo perceptions of matsutake elusiveness as something independent of human worlds.

By attending to the production of matsutake commodity elusiveness in contemporary Japan, and to the contradictions underpinning it, we see that the notion of elusiveness can both elide and reveal. This dynamic of 'hiding and revealing' is part of the elusiveness of matsutake elusiveness. Ethnographic attention to elusiveness requires that we hold on to both aspects of this valuation. Matsutake's status as a mediating commodity rests on an elusive duality. Just as the mushroom enables 'common people' to experience elite culture and participate in nationalist histories, it does so only through a brief experience of consumption under unpredictable, paradoxical, and uncontrollable circumstances. Similarly, just as the mushroom's charisma rests on its ability to link the urban and the rural, and the common and the elite, its status as 'elusive' shows up the contradictions inherent in its capacity to do so.

Acknowledgments

This chapter is the product of my collaboration with members of the Matsutake Worlds Research Group: Timothy Choy, Miyako Inoue, Michael Hathaway, Shiho Satsuka, Anna Tsing, and Elaine Gan. I am especially grateful to those in Japan who permitted me to interview them and otherwise supported my research. I also thank Jessica Cattelino, Hannah Landecker, Rachel Lee, Purnima Mankekar, Mariko Tamanoi, Akiko Takeyama, and Kathy Chetkovich for comments on earlier drafts. Funding for the fieldwork upon which this chapter is based was generously provided by the Terasaki Center for Japanese Studies and by the Toyota Foundation.

Lieba Faier is an Associate Professor of Geography at the University of California, Los Angeles. Her first book, *Intimate Encounters: Filipina Women and the Remaking of Rural Japan* (2009), is an ethnography of cultural encounters among Filipina migrants and their Japanese families and communities in rural Nagano. She is currently writing a second book, which examines ongoing efforts to fight human trafficking in Japan. She has published in *American Ethnologist*, *Cultural Anthropology*, *Annual Review of Anthropology*, *Transactions of the Institute of British Geographers*, and *Environment and Planning A*.

Notes

1. During peak season in Nagano, one of the primary regions where matsutake are found in Japan today, produce market auctions are usually limited to 100 kilos of matsutake per day. On Monday, 18 October 2010, a prodigious 334 kilos of local matsutake came into auction (Nippon Terebi 2010).
2. Unless otherwise indicated, all translations are my own.
3. In the 1930s, matsutake begin to assume a military nationalist flavor. During this period, tourism in Japan, which initially developed in large part to attract foreign visitors (especially Europeans and Americans) but later expanded to serve a domestic clientele, began to be framed in more explicitly nationalist terms and to focus increasingly on defining Japan in terms of its unique national character and culture (Leheny 2000).
4. By the time matsutake season begins in Japan in October, the importer had lamented, "a lot of consumers are tired of seeing matsutake."

References

Andoh, Elizabeth. 1995. "Fall Feasting; Japan." *New York Times*. 12 November. http://www.nytimes.com/1995/11/12/magazine/fall-feasting-japan.html.

Appadurai, Arjun. 1986. "Introduction: Commodities and the Politics of Value." *The Social Life of Things: Commodities in Cultural Perspective*, ed. Arjun Appadurai, 3–63. Cambridge: Cambridge University Press.

Appadurai, Arjun. 1988. "How to Make a National Cuisine: Cookbooks in Contemporary India." *Comparative Studies in Society and History* 30 (1): 3–24.

Arioka, Toshiyuki. 1997. *Matsutake*. Tokyo: Hosei University Press.

Arora, David. 1986. *Mushrooms Demystified: A Comprehensive Guide to the Fleshy Fungi*. 2nd ed. Berkeley: Ten Speed Press.

Asahi Shimbun. 1978. "Matsutake te ga demasen" [We can't afford matsutake]. *Asahi Shimbun*, 12 September.

Asahi Shimbun. 2010a. "Kaori no ōsama, omotomeyasuku" [The king of fragrance, much more affordable]. *Asahi Shimbun*, 20 October.

Asahi Shimbun. 2010b. "Matsutake sanwariyasu" [Matsutake, 30 percent off]. *Asahi Shimbun*, 4 October.

Benjamin, Walter. 1968. "The Work of Art in the Age of Mechanical Reproduction." *Illuminations: Essays and Reflections*, ed. Hannah Arendt; trans. Harry Zohn, 217–252. London: Jonathan Cape.

Bestor, Theodore C. 2001. "Supply-Side Sushi: Commodity, Market, and the Global City." *American Anthropologist* 103 (1): 76–95.

Boisard, Pierre. 2003. *Camembert: A National Myth*. Trans. Richard Miller. Berkeley: University of California Press.

Bourdieu, Pierre. 1984. *Distinction: A Social Critique of the Judgement of Taste*. Trans. Richard Nice. Cambridge, MA: Harvard University Press.

Caldwell, Melissa L. 2002. "The Taste of Nationalism: Food Politics in Postsocialist Moscow." *Ethnos* 67 (3): 295–319.

Caldwell, Melissa L. 2004. "Domesticating the French Fry: McDonald's and Consumerism in Moscow." *Journal of Consumer Culture* 4 (1): 5–26.

Callon, Michel, Cécile Méadel, and Vololona Rabeharisoa. 2002. "The Economy of Qualities." *Economy and Society* 31 (2): 194–217.

Chin, Elizabeth. 2001. *Purchasing Power: Black Kids and American Consumer Culture*. Minneapolis: University of Minnesota Press.

Choy, Timothy. 2011. *Ecologies of Comparison: An Ethnography of Endangerment in Hong Kong*. Durham, NC: Duke University Press.

Condry, Ian. 2006. *Hip-Hop Japan: Rap and the Paths of Cultural Globalization*. Durham, NC: Duke University Press.

Creighton, Millie. 1997. "Consuming Rural Japan: The Marketing of Tradition and Nostalgia in the Japanese Travel Industry." *Ethnology* 36 (3): 239–254.

Deloria, Philip J. 2004. *Indians in Unexpected Places*. Lawrence: University Press of Kansas.

DeSoucey, Michaela. 2010. "Gastronationalism: Food Traditions and Authenticity Politics in the European Union." *American Sociological Review* 75 (3): 432–455.

Faier, Lieba. 2009. *Intimate Encounters: Filipina Women and the Remaking of Rural Japan*. Berkeley: University of California Press.

Faier, Lieba. 2011. "Fungi, Trees, People, Nematodes, Beetles, and Weather: Ecologies of Vulnerability and Ecologies of Negotiation in Matsutake Commodity Exchange." *Environment and Planning A: Economy and Space* 43 (5): 1079–1097.

Foster, Robert J. 2005. "Commodity Futures: Labour, Love and Value." *Anthropology Today* 21 (4): 8–12.

Foster, Robert J. 2007. "The Work of the New Economy: Consumers, Brands, and Value Creation." *Cultural Anthropology* 22 (4): 707–731.

Fujokai. 1936. "Matsutake to shimeji no itadakikata" [Ways to enjoy having matsutake and shimeji]. *Fujokai* 4 (54).

Futai, Kazuyoshi. 2008. "Pine Wilt in Japan: From First Incidence to the Present." In Zhao et al. 2008, 5–12.

Ichinoe, Iseko. 1924. "Matsutake to kuri no oryōri hasshu" [Eight matsutake and chestnut recipes]. *Fujokai* 4 (30).

Ichige, Gorō. 1930. "Aki wo daihyō suru matsutake ryōri" [Matsutake cuisine that epitomizes autumn]. *Fujokai* 4 (42).

Inada, Shizue. 1939. "Aki no mikaku wo sosoru matsutake ryōri jūsshu" [Ten matsutake dishes that entice people into the taste of autumn]. *Fujin Kōron* 24 (10).

Ivy, Marilyn. 1995. *Discourses of the Vanishing: Modernity, Phantasm, Japan*. Chicago: University of Chicago Press.

Kamata, Naoto. 2008. "Integrated Pest Management of Pine Wilt Disease in Japan: Tactics and Strategies." In Zhao et al. 2008, 304–322.

Kamei, Makiko. 1938. "Fūmi yutakana matsutake ryōri jūsshu" [Ten flavorful matsutake dishes]. *Fujokai* 58 (4): 379–381.

Kelly, William W. 1986. "Rationalization and Nostalgia: Cultural Dynamics of New Middle-Class Japan." *American Ethnologist* 13 (4): 603–618.

Leheny, David. 2000. "'By Other Means': Tourism and Leisure as Politics in Pre-war Japan." *Social Science Japan Journal* 3 (2): 171–186.

Leitch, Alison. 2003. "Slow Food and the Politics of Pork Fat: Italian Food and European Identity." *Ethnos* 68 (4): 437–462.

Mankekar, Purnima. 2015. *Unsettling India: Affect, Temporality, Transnationality.* Durham, NC: Duke University Press.

Murakoso, Hitoshi, and Eiji Tanesaka. 2013. "A Historical Overview of the Matsutake Mushroom in the Heguri-Yama." *Memoirs of the Faculty of Agriculture of Kinki University* 46: 355–364.

Nippon Terebi. 2010. "Matsutake hōsaku nyūsu: Shinshū Ueda matsutake sanchi wo kinkyū shuzai! Nazo no daihōsaku" [Urgent reporting from a matsutake producing area (Ueda City, Nagano)! An unexpected bumper harvest]. *The Sunday Next*, 25 October.

Ohnuki-Tierney, Emiko. 1993. *Rice as Self: Japanese Identities through Time.* Princeton, NJ: Princeton University Press.

Ono, Kikuko. 1925. "Matsutake ryōri jūsshu" [Ten matsutake dishes]. *Ai no Izumi*, October.

Patico, Jennifer. 2002. "Chocolate and Cognac: Gifts and the Recognition of Social Worlds in Post-Soviet Russia." *Ethnos* 67 (3): 345–368.

Paxson, Heather. 2008. "Post-Pasteurian Cultures: The Microbiopolitics of Raw-Milk Cheese in the United States." *Cultural Anthropology* 23 (1): 15–47.

Saito, Haruo, and Gaku Mitsumata. 2008. "Bidding Customs and Habitat Improvement for Matsutake (*Tricholoma matsutake*) in Japan." *Economic Botany* 62 (3): 257–268.

Satsuka, Shiho. 2014. "The Satoyama Movement: Envisioning Multispecies Commons in Postindustrial Japan." *RCC Perspectives* 3: 87–94.

Shankar, Shalini. 2012. "Creating Model Consumers: Producing Ethnicity, Race, and Class in Asian American Advertising." *American Ethnologist* 39 (3): 578–591.

Sterngold, James. 1992. "Matsutake Madness Seizes the Japanese Every Autumn." *New York Times*, 11 November. http://www.nytimes.com/1992/11/11/garden/matsutake-madness-seizes-the-japanese-every-autumn.html.

Suzuki, Motoaki. 2010. "Waa! Matsutake nyokinyoki" [Wow! Matsutake popping up one after another]. *Asahi Shimbun*, 13 October.

Terrio, Susan J. 1996. "Crafting Grand Cru Chocolates in Contemporary France." *American Anthropologist* 98 (1): 67–79.

Trubek, Amy B. 2008. *The Taste of Place: A Cultural Journey into Terroir.* Berkeley: University of California Press.

Tsing, Anna. 2013. "Sorting Out Commodities: How Capitalist Value Is Made through Gifts." *HAU: Journal of Ethnographic Theory* 3 (1): 21–43.

Weiner, Annette B. 1992. *Inalienable Possessions: The Paradox of Keeping-While-Giving.* Berkeley: University of California Press.

Yamanaka, Katsuji. 2011. "Matsutake no rekishi to bunka" [The history and culture of matsutake]. *Shokuseikatsu* 105 (12): 26–32.

Yanagisako, Sylvia Junko. 2002. *Producing Culture and Capital: Family Firms in Italy.* Princeton, NJ: Princeton University Press.

Yomiuri Shimbun. 1953. "Matsutake daihōsaku, sengo saikō" [An enormous matsutake harvest, the most in the post-war period]. *Yomiuri Shimbun*, 5 October.

Yomiuri Shimbun. 2000. "Fukeiki … demo takane Ōyodo no ichiba ni hatsumatsutake, Nara" [High prices despite the recession, the first matsutake in Ōyodo market, Nara]. *Yomiuri Shimbun*, 3 October.

Yomiuri Shimbun. 2009. "Rekishi tsutaeru 'matsutake dōchū'" [A 'matsutake journey' that transmits history]. *Yomiuri Shimbun*, 5 October.

Zhao, Bo Guang, Kazuyoshi Futai, Jack R. Sutherland, and Yuko Takeuchi, eds. 2008. *Pine Wilt Disease*. Tokyo: Springer Japan.

Chapter 2

ELUSIVE FUNGUS?
Forms of Attraction in Multispecies World Making

Michael J. Hathaway

When anthropologists consider the 'more-than-human', they typically examine other species in relationship to humans. In contrast to the other chapters in this book that focus on the role of elusiveness, this chapter moves beyond a human center to explore the key companion term to elusiveness: attraction. The realm of attraction and elusiveness is based on different organisms' ability to detect and perceive others through their senses, and here I explore the sensorial worlds of less commonly considered species such as insects, plants, and fungi. In particular, I focus on the lives of mushrooms and how they participate in a wide range of sensorial engagements with a variety of organisms, including humans. Mushrooms such as the matsutake make themselves attractive, selectively engaging with others, and are themselves also attracted to other organisms.

References for this chapter begin on page 50.

I suggest that these two operations—elusiveness and attraction—occur simultaneously as species interact and communicate, evading some and seducing others. Although we often imagine elusiveness from a human-centered perspective, it is worthwhile to keep in mind that the notion of the elusive is itself relational. Being elusive is dependent on the notion of attraction: something cannot be elusive unless something else is attracted to it and trying to find it. Yet forms of attraction and elusiveness are quite diverse and cannot always be captured in human-centered terms. Something that is elusive to humans may seem quite accessible to other species. Indeed, attraction occurs in countless forms across biological kingdoms. For example, truffles' scents can penetrate a meter of soil and attract animals to dig it up, consume it, and deposit it elsewhere. Some tree fungi spores stick to woodpecker feathers and are carried to other trees after woodpeckers create new excavations. When we expand our interest in elusiveness beyond the human to fungal organisms like matsutake, we begin to see that they are both attractive and attracting through senses that we cannot easily perceive or understand, such as forms of chemical communication.

As I will show in this chapter, mushrooms are not just merely growing, or the objects of attraction for insects, birds, and mammals; they are also interpreting the world. Many of biologists' discoveries in this field of biosemiotics are quite new and surprising, and this chapter draws deeply on a critical social scientific reading of the biological literature. Such readings help reveal the diverse liveliness often hidden from view in accounts where humans are regarded as the only actor that matters, and mushrooms are regarded solely as commodities to be bought and sold. What happens when humans join in the search for the matsutake at the same time as many insects and other species that are attracted to these mushrooms? How does the mushroom itself engage its many interlocutors?

This chapter builds on the recent explosion of interest in the more-than-human across a number of fields, including history, geography, sociology, philosophy, and ethics (Birke 2003; Livingston and Puar 2011; Whatmore 2002). Work by anthropologists and others has added greatly to our understandings of animal agency, and even helped us question assumptions about how we might understand agency itself (Hathaway 2015; Johnson 2018; Kirksey and Helmreich 2010). The majority of such studies explore human relations with other fellow mammals, such as apes, dogs, and mice (Fuentes 2012; Haraway 2003, 2008; Moore and Kosut 2014). More-than-human studies are slowly expanding beyond a mammalian focus. For example, Hugh Raffles's (2010) *Insectopedia*, a masterful study of human-insect relations, adds to these conversations by considering beings often regarded as outside many people's normal 'moral circle'. Going beyond animals, a handful of scholars have begun to explore the worlds of bacteria, viruses, fungi, and plants (Besky and Padwe 2016; Hird and Roberts 2011; Lowe 2010; Myers 2015; Paxson 2008). The work of our collaborative project on the matsutake mushroom has often maintained a human-centered enterprise, but we have delved into some of ways these fungi play roles beyond their relations to humans (MWRG 2009a, 2009b; Tsing 2015). One of the many insights arising from studies of non-animals has been to reveal the continuing animal-centric perspectives among natural and social scientists about what constitutes agency and action, movement and sensing.

Yet as important as this work has been, Marianne Lien's (2015: 9) remarkable book, *Becoming Salmon*, argues that even when multispecies studies scholars go beyond the mammal, such studies are almost always "bilateral relations between human and single nonhuman species … rather than multispecies entanglements." In other words, even after years of social scientific work, anthropocentrism endures in a bilateral focus on relations between humans and one other species. Even in these bilateral analyses, humans often remain the central actors. Following Lien, I argue that stepping away from a human center and emphasizing the unique life-worlds created by relationships among different species provides new insights into multispecies entanglements. My focus on attraction and evasion in multispecies worlds helps us understand how different organisms interpret the world and interact with each other, even in the absence of humans, and challenges remaining tendencies toward anthropocentrism and human exceptionalism.

Attraction and Desire: Moving Beyond Human-Centered and Neo-Darwinist Accounts

Looking at attraction as a multispecies phenomenon expands recent social science scholarly interest in affect, desire, and charisma, terms commonly regarded as relating only to humans (Faier 2007; Rofel 2007). Disciplines such as psychology and biology tend to propose 'hard-wired' accounts of attraction between people or between people and other animals as generated through unconscious chemical signals, innate qualities, and ingrained predispositions (Fuentes 2012). In contrast, social scientists tend to denaturalize desire and affect, examining how they emerge in relations of power (Biehl and Locke 2010) and emphasizing their diversity, while focusing almost exclusively on human-centered questions of subject making. Cultural geographer Jamie Lorimer, one of the few social scientists to consider how humans become attracted to other species, proposes the concept of 'non-human charisma'. Lorimer (2007) argues that people such as birders, wildlife scientists, and environmentalists experience particular species as charismatic through repeated encounters (see also Parreñas 2012). We need to look elsewhere, however, for understanding forms of attraction that exist outside of a human center.

A focus on attraction helps build a language to explore multispecies world-building dynamics in part through a better understanding of ways of perceiving and making worlds that may vary significantly from human experience. My understanding of 'attraction' differs from that of natural scientists, who tend to view it as a manifestation of a biological sexual drive. I use the term to explore diverse forms of interest in other things, living and non-living. Attractions can be instinctual or learned (categories that are likely not mutually exclusive), mutual or one-sided, and obviously or ambiguously related to everyday survival. Attraction requires attunement through diverse forms, such as the ability to perceive and be interested in chemicals, odors, sounds, and sights. To live means to pursue attractions, and some species are simultaneously the objects of many forms of attraction as well as pursurers of others.

One path into the realm of multispecies attraction is through interspecies communication, and Eduardo Kohn's (2013) *How Forests Think* is exemplary in this regard. Drawing on the path-breaking work of polymath C. S. Peirce, who argues that semiosis (creating and interpreting signals from others) is not limited to humankind but is instead an activity of all living beings, Kohn extends Pierce's thinking into new areas. Within the social sciences, opening up semiotics to the more-than-human represents a serious challenge to entrenched beliefs in human exceptionalism (Plumwood 2002).

One of Kohn's critical interventions is to focus on other living beings in terms of their specificity and not just as generic non-humans, a common approach reinscribed by the term itself even in work that decenters a human monopoly on agency (e.g., Bruno Latour's 'actant' or Jane Bennett's 'vibrant matter'). Whereas Kohn focuses more on qualities of mind (i.e., how forests think), I explore how different bodies inhabit the world and create worlds through their liveliness. I suggest that this move accomplishes two important goals: first, it recognizes the diversity of ways of being in the world, of which humans are just one of many; second, it situates human capacities, recognizing and specifying them. One of the pitfalls of not noticing the role of human perception is that we tend to conflate it with reality itself.

Exploring such diverse and multispecies forms of attraction takes us beyond an exclusive concern with human-based senses. When it comes to the human species, anthropologists have shown us how diverse languages and ontologies structure groups' experiences of the world in radically different ways (Holbraad et al. 2014; Mol 2002), and these may change dramatically over time. Since the eighteenth century, for example, Westerners have moved away from the use of smell and touch in everyday life to privilege vision above other senses (Haraway 1991). Recently, scholars have considered other senses, such as taste and sound (Inoue 2006; Pink 2010), but these studies have rarely explored how such senses are shaped by the human *umwelt* or have compared our senses with those of other species. Taking more-than-human senses seriously helps us better imagine life-worlds of other beings for whom vision is not nearly as central or whose qualities of sight are radically different from our own.

Umwelt: Insights into Sensorial Worlds

To explore attraction and sensorial worlds in terms of the specific qualities of particular beings, one of the most promising approaches is the work of Jakob von Uexküll (1864–1944), a Baltic German biologist who is mentioned but not highly elaborated on by Kohn. In 1909, Uexküll introduced the concept of the *umwelt*—the world as it is sensorially experienced by particular organisms (see Uexküll [1909] 2010). *Umwelt* is the subjective environment that each species perceives and creates, which is shaped by its sensory apparatus—its specific capacities to perceive and interpret smell; see ultraviolet, color, or black and white; hear different bands of the sound spectrum or detect other vibrations. Uexküll asks, how does a cow or a jellyfish perceive the world? Many of our scientific understandings of the dances of bees, the songs of birds and whales, and the use of insect pheromones have been influenced by Uexküll's sustained curiosity.

My use of the *umwelt* draws on Uexküll but takes his work in a new direction. First, I see an organism's *umwelt* as shaped through its encounters in the particular place in which it dwells. Here I follow anthropologist Tim Ingold (2000), who understands *umwelt* not as a fixed sensorial apparatus but as an active process of world making by different species (see also Magnus 2014; Maran et al. 2012), and Eva Hayward (2010), who shows us how sensing is related to how organisms make sense of the world. The *umwelt* is not "a touchable and tangible category, but rather an array of subjective and perceptive elements" (Maran et al. 2012: 12). My use of the *umwelt* goes against the main invocation of Uexküll's concept in the social sciences and humanities—the frequently invoked story of the woodland tick and its limited sensorial universe (Agamben 2003; Deleuze and Guattari 1987). Unfortunately, this story gives the impression of the *umwelt* as fixed and impoverished. Instead, I suggest that organisms actively shape their *umwelt* through their continual engagement with the world as they experience and live it.

Second, my reading of *umwelt* stresses how organisms come into being in relationship to each other. This use may differ from Uexküll's famous metaphor of autonomous soap bubbles to describe the existence of different species trapped in their own perceptual universes, their own self-contained worlds. We know, however, that species continually engage with other species through attraction and evasion. For organisms to survive as a species, they must connect themselves with the lives of others—as predators, prey, and parasites; as flowers and pollinators; as sources of spores and agents of the spores' dispersal; as co-inhabitants. In other words, Uexküll pays insufficient attention to the fact that many *umwelten* actually overlap and intersect. This oversight is most obvious when we think of animals that are attracted to living food to survive, such as bobcats hunting for squirrels. In this way, the bobcat's ability to pounce on a squirrel depends on its ability to imagine the squirrel's *umwelt* (i.e., to approach from downwind and in ways not visible to the squirrel). Although they do so in very different ways, plants must also perform their daily quest to meet their needs (water, soil nutrients, sunlight) in dynamic environments to which they must adjust and respond. In turn, many flowering plants utterly depend on attracting insects in order to create fruit and spread seeds and thus reproduce over the generations. Hence, the act of pollination is an intersection of *umwelten*.

We can turn the *umwelt* back on ourselves as Uexküll's notion alerts us that our experiences and understandings are shaped by our human bodies. Thus, many multispecies engagements are elusive to us in part because we lack a capacity to sense them: they are too small to see or use ultraviolet colors we cannot detect; their sounds are at higher or lower frequencies than we can hear; and they are mediated by odors that we cannot smell. For example, many terms that seem like scientific universals are in fact based on the human *umwelt*: 'microscopic' refers to objects too small for the human eye, and 'ultrasonic' refers to sounds higher than the human spectrum of hearing. As Jamie Lorimer (n.d.) notes, acknowledging the human *umwelt* quickly reveals how limited our capacities are: "Unlike most terrestrial mammals that communicate with pheromones, we depend on vision and privilege visual knowledge. Human sensory organs make use of small portions of the electromagnetic, acoustic, and olfactory spectra for perception and communication."

Thus, all species, including humans, use their senses to actively interpret the universe, and these interpretations play a role in making that universe. Awareness of our own limited abilities, along with noting the capabilities of other species (e.g., whales and bats), helped us develop technologies such as radar and sonar. Instead of conflating what we notice via our limited senses with the world itself, we become aware of how our senses shape the contours of the worlds we make.

Yet even if social scientists build on Uexküll's insights into and curiosity about other species, they should not merely reproduce scientific knowledge but should show how such knowledge is situated and rooted in particular theoretical frameworks (Haraway 1988, 1989; Strum and Fedigan 2000). Under the reigning neo-Darwinist biological paradigm, relationships are understood within a relentless cost-benefit framework, a position critiqued by a few biologists under the banner of 'liberation biology' (Rose 1982). As of yet, relatively few social scientists challenge such biological frameworks with the notable exception of 'queer ecology' (Mortimer-Sandilands and Erickson 2010), which explicitly questions assumptions about binary gender formations and heteronormative behaviors (see also Hird and Roberts 2011).

Motivated by this scholarship, an essay by Carla Hustak and Natasha Myers (2012) reveals some dominant scientific assumptions and makes a provocative argument for the role of attraction and desire outside of the logic of cost-benefit calculation. They show how some orchids attract insects with scents strongly reminiscent of insect sex pheromones. Botanists describe these encounters as orchids tricking insects to mate with their flowers: the insect loses by failing to reproduce, and the orchid wins by spreading its spores. Hustak and Myers, however, argue against subjecting these relations to a maximizing logic and argue for considering interspecies attractions between orchids and insects as "creative, improvisational, and fleeting practices through which plants and insects *involve* themselves in one another's lives" (ibid.: 77). Hustak and Myers's position, which resonates strongly with my own, entertains the possibility of non-human lives not totally dictated by a kind of utilitarianism that is always optimizing foraging efficiency or maximizing progeny (i.e., better 'fitness') but is instead inflected by kinds of attraction and desire on the part of the fly and orchid that we may not necessarily notice or understand. Without anthropomorphizing (imparting human motives onto other species), we might inquire into divergent forms of attraction across diverse forms of life.

In the last half of the chapter, I draw on the biological literature through a critical social scientific analytic framework to reveal implicit frameworks, such as neo-Darwinism and animal centrism. I begin with an exploration of forms of attraction through the *umwelt* of plant worlds and conclude with an exposition of how fungi live in worlds of attractiveness and elusiveness, drawing on the relatively scant mycological literature currently available that is specific to matsutake liveliness. I first discuss plants because botanists have conducted far more studies on vegetal capacities than mycologists have done on fungal capacities. I show that once we move beyond the human, we can move beyond familiar senses, such as seeing, smelling, and hearing, into less well-known terrain, such as detecting heat, perceiving ultraviolet light, or using echolocation. Moving beyond the animal, we encounter even lesser known senses such as cellular or photonic (the generation, detection, and manipulation of light) capacities.

Rethinking Attraction beyond the Human: Plants

What does research on botanical communication and sensorial capacities reveal that might shed light on fungi and help broaden our understanding of multispecies becoming? Plants, like fungi, are often regarded as mute and sessile, particularly when compared to animals that can walk, run, jump, and fly (Chamovitz 2012). Based on our human experience of time, they do not seem to move—only after some passage of time do we tend to notice plants growing upward and outward.

Even the notably non-anthropocentric field of biosemiotics (communication among non-humans, a field with roots in Uexküll's scholarship) has historically been exclusively concerned with animals—the realm of plant semiotics has only recently been explored (Kull 2000). The Austrian philosopher Günther Witzany

(2006: 169), likely the world's most prolific philosopher of biosemiotics, argues that plants use chemicals "as signals, messenger substances, information carriers and memory medium in either solid, liquid or gaseous form." He finds that plants actively interpret signs, both biotic and abiotic. Witzany (2012: 1) suggests they follow three kinds of rules, similar to human communication: participants interpret meanings through "syntactic (combinatorial), pragmatic (context dependent) and semantic (content-specific)" contexts. Although a philosopher, Witzany works almost exclusively with peer-reviewed scientific papers. He explores communication at three levels: within the organism's body, between plants of the same and different species, and between plants and animals. He argues that air is often thick with communication, much of which is extraneous to the plants' livelihood, what we call 'noise'. Plants have an extensive chemical 'vocabulary': there are more than 20 different types of chemical communication, and plant roots can produce more than 100,000 different chemicals (Witzany 2006: 170). Each of these chemicals can be a form of communication, and each chemical reacts with others to create novel signals.

Insect-wounded plants create one form of chemical communication. These chemical messages are received by nearby plants, even different species, which in turn secrete chemicals that make themselves less appetizing to insects. Other plants communicate across kingdoms and produce chemicals detected by animals, attracting the predators of these attacking insects. Such signals are not simple 'all or nothing' alerts, but differ in proportion to the attack. As well, these signals distinguish between the types of attack. Plants cut with a knife do not respond with the same message as those eaten by herbivores, where it seems that plants can detect chemicals in different herbivores' saliva. Producing protective chemicals would not help plants cut with a knife. Such effort would be without purpose, and somehow the plants recognize the difference (Felton and Tumlinson 2008). Plants expend a lot of energy in creating chemicals (both in producing an alert and in response to others' alerts), and therefore it is vital that they correctly interpret airborne signals (Witzany 2006: 170). If plants produced protective chemicals in response to every false alert, they would soon use up their own internal resources and perish. This insight—that plants actively interpret the world—challenges the typical assumption that plants are basically passive and inert. They attend to important signals and produce their own: plants are almost constantly using chemicals to both elude and attract others.

Chemicals are not the only way that plants experience the world, and from a perspective that follows Uexküll's refusal of anthropocentrism, the botanist Daniel Chamovitz (2012) argues that biologists have tended to define senses in overly animal-centric terms. For example, he points out that the most common scientific understanding of sight conforms to *Merriam-Webster*'s definition, that is, "the physical sense by which light stimuli received by the eye are interpreted by the brain and constructed into a representation" (ibid.: 10). Instead, he asks, how we might understand 'sight' for organisms like plants that have neither eyes nor brains?

Thus, by expanding our notions of what counts as seeing, as well as other senses such as hearing and tasting, we can learn more about how plants perceive the world. While most people would assume that plants' perceptual ability is much less than our own, it might come as some surprise to hear that humans see with just two main forms of photoreceptors (rods and cones), whereas plants have at least eleven kinds of photoreceptors. Although we do not yet understand how plants visually interpret the world, they likely have abilities far beyond our own, and understanding such abilities will take many creative experiments.

This scientific interest in botanical agency is part of a broader recognition that is having increasing social and political impact. Recently, the rights of plants have been considered in legal terms. One of the most dramatic cases occurred in Switzerland: in the 1990s the constitution was changed to acknowledge 'plant dignity'. Some scientists have dismissed this as quackery; one biologist suggests it represents a path 'to absurd land' (Lev-Yadun 2008). Yet Florianne Koechlin (2009: 78), a defender of the Swiss statement, argues that such so-called rigorous scientific positions may actually be based on shaky foundations:

> We do not know if plants are capable of subjective sensation. There is no scientific proof that plants feel pain. But it is also quite clear that we cannot simply rule this out. There is circumstantial evidence for this, although not a complete chain of evidence. However, claims that plants have no subjective sensations are as speculative as the opposite. We simply do not know. We cannot deny with certainty that plants lack an ability to actively perceive. Thus far, plant abilities to perceive their environment has been widely underestimated- But what could be the consequences of these new findings? How should we approach this situation of 'not knowing'?

Lev-Yadun's dismissive position may be quite widespread, where in the name of scientific rigor there is a tendency to dismiss the capacities of other living beings. Natasha Myers's works with botanists on questions of vegetal capacities shows a similar concern, and this history of mainstream scientific antagonism toward vegetal and fungal capacities has seriously hampered efforts by botanists and mycologists.

Fungi

Learning about plants can also help guide our studies of fungal attraction. Many Western mycologists have assumed that fungal chemicals are mainly used to deter potential predators (Mithöfer and Boland 2012), but truffles are a prominent exception, for their lives are obviously and utterly dependent on attracting mammals and insects to spread their spores. Truffles create odors that only certain animals can detect. In some cases, these relations are highly specific: mushroom spores may depend heavily on one animal species and yet be destroyed by a closely related species (Maser et al. 2008). Thus, it makes sense that mushrooms' strategies to disperse spores often combine simultaneous efforts to elude some harmful species and attract other helpful species. But scientists still know relatively little about the fate of animal-consumed spores, for example, which fungal spores can survive through the digestive system of which animals, whether mammals, birds, or insects.

Biologists know, however, that some truffles successfully rely almost completely on one species of animal to disperse their spores—flying squirrels (Maser et al. 2008). Ripe truffles release a scent chemically reminiscent of the sex pheromones of flying squirrels, which subsequently dig them up. Spores stay viable as they move through the squirrel's digestive system and are deposited elsewhere via its feces (Pyare and Longland 2002). Some of these squirrels, in turn, depend almost completely on truffles and eat little else.

Other truffles reproduce by attracting insects. In the case of the tiny beetle *Leiodes cinnamomea*, only the female perceives the truffle. Females burrow underground, and the males, in turn, are attracted to the females' scent. Beetles of both sexes consume their subterranean feast, spreading spores as they defecate (Hochberg et al. 2003). Thus, in this case the beetles' odor *umwelt* is gender-specific. Again, without their insect transport system the truffles could die off.

Besides its function as a source of food, why might other species be attracted to fungi? They might seek it as a shelter, a place to raise young, or a hunting ground. Conversely, fungi themselves exert some forces of attraction, such as when fungal-infected 'zombie' caterpillars produce yet unknown volatiles that attract mosquitoes that bite them and are subsequently turned into fungal agents (George et al. 2013). Fungi can also take over plant morphology and behavior, creating 'zombie plants' by transforming plants' bodies through fungal action. For example, fungi can force plants to flower earlier or to produce fungal spores rather than pollen (Jennersten 1988). Mushrooms attract insects in a number of ways through shapes, odors, color, and phosphorescence. For many years, fungal phosphorescence was assumed to be a mere metabolic byproduct, but in the 1980s mycologists found that fungi modulated this glow and that it did attract spore-carrying insects at night (Sivinski 1981). Why did this take so long to understand? One reason, mentioned earlier, was that many Western mycologists assumed that fungi acted only to defend themselves against insect attack instead of attracting certain insects. Second, this discovery also remained elusive in part because of human's sensory affordances, specifically our poor ability to see at night. Thus, what we understand as elusive is in fact elusive to us due to our perceptual abilities and entrained capacities, and yet we tend to project our human *umwelt* onto other species or even to conflate our perception with reality itself.

Attraction, often seen as a kind of one-to-one relation between two individuals, may be far more complex. In the case of certain mushrooms, mycologists have typically understood attraction as a continual effort, such as the constant emission of a chemical, but sometimes efforts at attraction are created quickly in response to specific actions and require finely honed abilities of discernment. For example, when certain mushrooms are attacked by a particular insect, the mushroom emits a chemical signal to a specific predator of that insect. In order to send such a signal, the mushroom has to properly identify the insect that is eating it, for predatory insects will only eat certain prey. Thus, the mushroom must be able to tell the difference between, say, a beetle that has a predator that the mushroom can call for assistance and a slug that does not have any predator that the mushroom can call (Witzany 2012). This sophisticated arrangement is initiated by the mushroom and relies on a mutual attunement between the mushroom and the predator. Even among animals that we tend to imbue with more active abilities to perceive the world and act in their own interest, we rarely find instances like this where the prey calls a third species to predate upon its attacker.

This kind of interspecies—or inter-kingdom—communication between animals and fungus, like many forms of communication, has likely evolved over long periods of time. Both insects and mushrooms learn to adjust to each other's presence. They evolve new sensory abilities and create new means to communicate over long distances, such as through producing, emitting, detecting, and responding to novel chemicals.

Multispecies Relationships Form through Entangled Dynamics of Attraction

What can attuning ourselves to these different forms of plant and fungal attraction teach us about cosmopolitan multispecies worlds? What happens when humans join in the search for the matsutake at the same time as many insects and other species that are attracted to these mushrooms, and how might the mushroom itself engage with these interlocutors?

Experienced mushroom hunters often find a mushroom too late: the insects have beat them to it, and the stem and gills are riddled with tunnels or the upper cap is chewed up. Although matsutake emits powerful odors, the few humans who find matsutake through smell rarely detect them from far: our capacity to smell pales in comparison to many insects and other mammals.

For humans, matsutake seem to pop up and disappear quickly, but other species may experience such time frames differently. In human time, the life span of a mushroom is quick, two weeks out of an average human lifetime of thousands of weeks; yet for insects, two weeks might be longer than their own lives.

Finding a single mushroom might be all that insects need to ensure the survival of their progeny, whereas human market hunters often aim to collect hundreds or thousands of mushrooms each year. They are trying to make a living, and their income is as dependent on successfully finding mushrooms as it is on the vagaries of the fluctuating value of their harvest. If humans miss when matsutake fruit in an area, they may hope that their timing is better the next year, or if their desire is strong enough, they may travel to colder or warmer climates or hunt in higher or lower elevations, depending on whether fruiting is triggered more by heat or cold. Plenty of hunters have discovered mushrooms just hours after other people, deer, pigs, birds, or slugs have already found them. When humans pick for their own consumption, those matsutake partially eaten by vertebrates can often be salvaged, but for the majority of hunters who pick to sell, such mushrooms are worth little or nothing. Staying alive or even thriving in a multispecies world always demands an attunement to others' *umwelten*, continually attracting some and eluding others.

Although matsutake may appear to humans to be elusive, paying attention to the entangled forms of attraction that shape their life-worlds offers a different picture. A wide range of species, not only humans and other mammals but also insects, are attracted to and capable of sensing the mushrooms. Some insects can detect some of the chemicals produced by mushrooms, even when hidden out of sight. In other words, such chemicals, whatever the mushrooms' 'intention', can become a form of attraction. More surprisingly, we might learn more about how mushrooms, like plants, are also interpreting the world, listening for signals from others, reacting based on communications from others of the same or different species. Witzany (2012) argues that fungi—whether communicating at the intra-cellular level to manage their own growth and defense between other members of the same species and with other species—must interpret chemical-born messages, ignoring some while responding to others based on context-dependent clues. They also create their own signals and send them through a mycelial network, what some call the 'wood wide web', as well as diffusing them into the atmosphere and soil matrix. Mushrooms, therefore, are not merely passive objects, elusive and lying in wait for active animals to discover them. Instead, mushrooms, like animals and plants, are semiotic beings, creating signals and interpreting them from others. Acts of semiosis take place in these and other more-than-human organisms that are also seeking food and mates, motivated by a broad and dynamic field of attractions.

Conclusion

A close examination of matsutake mushrooms and their fungi cousins shows that organisms are always shaped by their relations with multiple species at once. Such effects are difficult to parse out. Social scientists can pay attention to these dynamics, but it can be hard to draw on natural scientific research to evaluate these multiple relations because experiments using the conventional scientific method usually isolate one relationship at a time. In the case of matsutake and humans, insects are the other main player that mediates their relationship, but there are many other players, most of whom I have not mentioned in this chapter. Some insect species have co-evolved with particular mushrooms, and their own *umwelt* is shaped by this relationship. They have learned to attune themselves to specific mushrooms and their chemical repertoire, and the same can be said for the mushroom itself. In turn, these intimate relationships powerfully influence how humans find, amass, and transport matsutake throughout the entire transnational commodity chain. As insects seek out mushrooms, bury themselves in their flesh, and use mushrooms as a breeding ground, humans are compelled to attempt to exclude the insects or to restrict their progress with the use of refrigeration, which slows down insect metabolism and activity.

I have shown how social scientists might selectively embrace the insights of Jakob von Uexküll to open up multispecies ontologies and gain better understanding of how our own *umwelt* as humans shapes our understandings of, and encounters with, the world. Yet just as we try to acknowledge the role that theoretical frameworks play in shaping knowledge and inflecting claims among our fellow social scientists, I suggest we do the same when exploring the rich insights generated by natural scientific studies. As shown in this chapter, strong tendencies remain (even among biologists) to evaluate diverse forms of life against an animal standard, and to regard non-animals as having diminished agency and capacities. We have seen that even though scientific work on vegetal and fungal capacities has been hampered through fear of being considered unscientific, we nonetheless are starting to learn more about the surprising ways in which fungi engage the world through their own *umwelt*. We can start to appreciate more and more how matsutake—in part through ejecting its spores into powerful winds and spreading across the planet, and in part through human-mediated commodity circuits—has become a transnational organism. It fosters and knits together relations in pine, oak, and chinquapin forests and creates a unique chemical bouquet that attracts a range of animals to seek it out as it shapes worlds, with and beyond human efforts at its management. Exploring matsutake's *umwelt* and how it creates these relations is one step toward understanding truly multispecies world making.

Acknowledgments

This work was supported by the Social Sciences and Humanities Research Council, Social Science Research Council, American Council of Learned Societies, and Simon Fraser University. Earlier versions of this chapter were presented by the Matsutake Worlds Research Group's panel at the 2012 American Anthropological Association conference in San Francisco and in two talks at Yale University in 2014 and 2015: an Agrarian Studies Colloquium and a workshop for Michael Dove. I would like to extend my thanks to Mel Johnson, Karen Hebert, Kathleen Millar, Kathy White, Juliet Erazo, Regis Groleau, Yuan Wei, and Leslie Walker Williams for their help in thinking about these ideas and in improving the writing.

Michael J. Hathaway is an Associate Professor of Cultural Anthropology at Simon Fraser University in Vancouver, British Columbia. His first book, *Environmental Winds: Making the Global in Southwest China* (2013), won the Cecil B. Currey Book Prize for the best book on development studies. It explores how environmentalism was refashioned in China, not only by conservationists, but also by rural villagers and even animals. This chapter draws from his second major project, with the Matsutake Worlds Research Group, which examines the global commodity chain of the matsutake, one of the world's most expensive mushrooms, following it from the highlands of the Tibetan Plateau to the markets of urban Japan.

References

Agamben, Giorgio. 2003. *The Open: Man and Animal.* Trans. Kevin Attell. Stanford, CA: Stanford University Press.

Besky, Sarah, and Jonathan Padwe. 2016. "Placing Plants in Territory." *Environment and Society: Advances in Research* 7: 9–28.

Biehl, João, and Peter Locke. 2010. "Deleuze and the Anthropology of Becoming." *Current Anthropology* 51 (3): 317–351.

Birke, Lynda. 2003. "Who—or What—Are the Rats (and Mice) in the Laboratory." *Society & Animals* 11 (3): 207–224.

Chamovitz, Daniel. 2012. *What a Plant Knows: A Field Guide to the Senses of Your Garden—and Beyond.* London: Oneworld Publications.

Deleuze, Gilles, and Félix Guattari. 1987. *A Thousand Plateaus: Capitalism and Schizophrenia.* Trans. Brian Massumi. Minneapolis: University of Minnesota Press.

Faier, Lieba. 2007. "Filipina Migrants in Rural Japan and Their Professions of Love." *American Ethnologist* 34 (1): 148–162.

Felton, Gary W., and James H. Tumlinson. 2008. "Plant-Insect Dialogs: Complex Interactions at the Plant-Insect Interface." *Current Opinion in Plant Biology* 11 (4): 457–463.

Fuentes, Agustin. 2012. "Ethnoprimatology and the Anthropology of the Human-Primate Interface." *Annual Review of Anthropology* 41: 101–117.

George, Justin, Nina E. Jenkins, Simon Blanford, Matthew B. Thomas, and Thomas C. Baker. 2013. "Malaria Mosquitoes Attracted by Fatal Fungus." *PLoS One* 8 (5): e62632. doi.org/10.1371/journal.pone.0062632.

Haraway, Donna. 1988. "Situated Knowledges: The Science Question in Feminism and the Privilege of Partial Perspective." *Feminist Studies* (14) 3: 575–599.

Haraway, Donna. 1989. *Primate Visions: Gender, Race, and Nature in the World of Modern Science*. New York: Routledge.

Haraway, Donna. 1991. *Simians, Cyborgs, and Women: The Reinvention of Nature*. New York: Routledge.

Haraway, Donna. 2003. *The Companion Species Manifesto: Dogs, People, and Significant Otherness*. Chicago: Prickly Paradigm Press.

Haraway, Donna. 2008. *When Species Meet*. Minneapolis: University of Minnesota Press.

Hathaway, Michael J. 2015. "Wild Elephants as Actors in the Anthropocene." In *Animals in the Anthropocene: Critical Perspectives on Non-Human Futures*, ed. Human Animal Research Network Editorial Collective, 221–242. Sydney: Sydney University Press.

Hayward, Eva. 2010. "Fingeryeyes: Impressions of Cup Corals." *Cultural Anthropology* 25 (4): 577–599.

Hird, Myra J., and Celia Roberts. 2011. "Feminism Theorises the Nonhuman." *Feminist Theory* 12 (2): 109–117.

Hochberg, Michael E., Guillaume Bertault, Karine Poitrineau, and Arne Janssen. 2003. "Olfactory Orientation of the Truffle Beetle, *Leiodes cinnamomea*." *Entomologia Experimentalis et Applicata* 109 (2): 147–153.

Holbraad, Martin, Morten Axel Pedersen, and Eduardo Viveiros de Castro. 2014. "The Politics of Ontology: Anthropological Positions." *Cultural Anthropology*. 13 January. https://culanth.org/fieldsights/462-the-politics-of-ontology-anthropological-positions.

Hustak, Carla, and Natasha Myers. 2012. "Involutionary Momentum: Affective Ecologies and the Sciences of Plant/Insect Encounters." *differences* 23 (3): 74–118.

Ingold, Tim. 2000. *The Perception of the Environment: Essays on Livelihood, Dwelling and Skill*. London: Routledge.

Inoue, Miyako. 2006. *Vicarious Language: Gender and Linguistic Modernity in Japan*. Berkeley: University of California Press.

Jennersten, Ola. 1988. "Insect Dispersal of Fungal Disease: Effects of Ustilago Infection on Pollinator Attraction in Viscaria vulgaris." *Oikos* 51 (2): 163–170.

Johnson, Melissa A. 2018. *Becoming Creole: Nature and Race in Belize*. New Brunswick, NJ: Rutgers University Press.

Kirksey, S. Eben, and Stefan Helmreich. 2010. "The Emergence of Multispecies Ethnography." *Cultural Anthropology* 25 (4): 545–576.

Koechlin, Florianne. 2009. "The Dignity of Plants." *Plant Signaling & Behavior* 4 (1): 78–79.

Kohn, Eduardo. 2013. *How Forests Think: Toward an Anthropology beyond the Human*. Berkeley: University of California Press.

Kull, Kalevi. 2000. "An Introduction to Phytosemiotics: Semiotic Botany and Vegetative Sign Systems." *Sign Systems Studies* 28: 326–350.

Lev-Yadun, Simcha. 2008. "Bioethics: On the Road to Absurd Land." *Plant Signaling & Behavior* 3 (8): 612.

Lien, Marianne E. 2015. *Becoming Salmon: Aquaculture and the Domestication of a Fish*. Berkeley: University of California Press.

Livingston, Julie, and Jasbir K. Puar. 2011. "Interspecies." *Social Text* 29 (1): 3–14.

Lorimer, Jamie. 2007. "Nonhuman Charisma." *Environment and Planning D: Society and Space* 25 (5): 911–932.

Lorimer, Jamie. n.d. "Charisma." Multispecies Salon. http://www.multispeciessalon.org/charisma/.

Lowe, Celia. 2010. "Viral Clouds: Becoming H5N1 in Indonesia." *Cultural Anthropology* 25 (4): 625–649.

Magnus, Riin. 2014. "The Function, Formation and Development of Signs in the Guide Dog Team's Work." *Biosemiotics* 7 (3): 447–463.

Maran, Timo, Dario Martinelli, and Aleksei Turovski, eds. 2012. *Readings in Zoosemiotics*. Berlin: De Gruyter Mouton.

Maser, Chris, Andrew W. Claridge, and James M. Trappe. 2008. *Trees, Truffles, and Beasts: How Forests Function*. New Brunswick, NJ: Rutgers University Press.

Mithöfer, Axel, and Wilhelm Boland. 2012. "Plant Defense against Herbivores: Chemical Aspects." *Annual Review of Plant Biology* 63: 431–450.

Mol, Annemarie. 2002. *The Body Multiple: Ontology in Medical Practice*. Durham, NC: Duke University Press.

Moore, Lisa Jean, and Mary Kosut. 2014. "Among the Colony: Ethnographic Fieldwork, Urban Bees and Intra-species Mindfulness." *Ethnography* 15 (4): 516–539.

Mortimer-Sandilands, Catriona, and Bruce Erickson, eds. 2010. *Queer Ecologies: Sex, Nature, Politics, Desire*. Bloomington: Indiana University Press.

MWRG (Matsutake Worlds Research Group). 2009a. "A New Form of Collaboration in Cultural Anthropology: Matsutake Worlds." *American Ethnologist* 36 (2): 380–403.

MWRG (Matsutake Worlds Research Group). 2009b. "Strong Collaboration as a Method for Multi-sited Ethnography: On Mycorrhizal Relations." In *Multi-sited Ethnography: Theory, Praxis and Locality in Contemporary Research*, ed. Mark-Anthony Falzon, 197–214. London: Ashgate.

Myers, Natasha. 2015. "Conversations on Plant Sensing: Notes from the Field." *NatureCulture* 3: 35–66.

Parreñas, Rheana "Juno" Salazar. 2012. "Producing Affect: Transnational Volunteerism in a Malaysian Orangutan Rehabilitation Center." *American Ethnologist* 39 (4): 673–687.

Paxson, Heather. 2008. "Post-Pasteurian Cultures: The Microbiopolitics of Raw-Milk Cheese in the United States." *Cultural Anthropology* 23 (1): 15–47.

Pink, Sarah. 2010. "The Future of Sensory Anthropology/the Anthropology of the Senses." *Social Anthropology* 18 (3): 331–340.

Plumwood, Val. 2002. *Environmental Culture: The Ecological Crisis of Reason*. London: Routledge.

Pyare, Sanjay, and William S. Longland. 2002. "Interrelationships among Northern Flying Squirrels, Truffles, and Microhabitat Structure in Sierra Nevada Old-Growth Habitat." *Canadian Journal of Forest Research* 32 (6): 1016–1024.

Raffles, Hugh. 2010. *Insectopedia*. New York: Pantheon Books.

Rofel, Lisa. 2007. *Desiring China: Experiments in Neoliberalism, Sexuality, and Public Culture*. Durham, NC: Duke University Press.

Rose, Steven, ed. 1982. *Towards a Liberation Biology: The Dialectics of Biology Group*. New York: Schocken Books.

Sivinski, John. 1981. "Arthropods Attracted to Luminous Fungi." *Psyche* 88 (3–4): 383–390.

Strum, Shirley C., and Linda Marie Fedigan, eds. 2000. *Primate Encounters: Models of Science, Gender, and Society*. Chicago: University of Chicago Press.

Tsing, Anna L. 2015. *The Mushroom at the End of the World: On the Possibility of Life in Capitalist Ruins*. Princeton, NJ: Princeton University Press.

Uexküll, Jakob von. (1909) 2010. *A Foray into the Worlds of Animals and Humans with a Theory of Meaning*. Trans. Joseph D. O'Neil. Minneapolis: University of Minnesota Press. Originally published in German as *Umwelt und Innenwelt der Tiere*.

Whatmore, Sarah. 2002. *Hybrid Geographies: Natures Cultures Spaces*. London: Sage.

Witzany, Günther. 2006. "Plant Communication from Biosemiotic Perspective: Differences in Abiotic and Biotic Signal Perception Determine Content Arrangement of Response Behavior. Context Determines Meaning of Meta-, Inter- and Intraorganismic Plant Signaling." *Plant Signaling & Behavior* 1 (4): 169–178.

Witzany, Günther. 2012. "Introduction: Key Levels of Biocommunication in Fungi." In *Biocommunication of Fungi*, ed. Günther Witzany, 1–18. Dordrecht: Springer.

Chapter 3

TENDING TO SUSPENSION

Abstraction and Apparatuses of Atmospheric Attunement in Matsutake Worlds

Timothy Choy

How might one conduct an anthropology of atmospheres and atmospheric subjects in atmospheric terms? What methods and capacities must one develop to work with objects that elude particular habitual knowledge and sensory practices? What does attention to atmospherics tell us about matsutake elusiveness, and what do mycological attunements tell us about elusiveness more generally?

Consider the odor of the matsutake, that coveted mushroom's most distinctive—and hard to characterize—feature. People wax poetic about the fragrance. Famed mushroom hunter David Arora (1986: 191) offers a mash of incongruous associations to evoke it: "'Spicy but a little bit foul' ... a provocative compromise

Notes for this chapter begin on page 74.

between 'red hots' and dirty socks." Matsutake's taste is as distinctive as its odor, "an incredible and complex flavor you won't ever forget—even though you won't be able to adequately describe it to anyone" (Volk 2000).

Alongside poetics are negatives. During preliminary research on the semiotics of smell in Japan, Miyako Inoue became captivated by what she calls the negativity of matsutake olfactory aesthetics among people she knows. By this, Inoue means to call attention to the moments and ways people disclaim the capacity to discern the distinctiveness of matsutake smell. Rather than the rapturous glottal explosion "Ah!" that prefaces the typical reply of matsutake lovers ("Ah! The smell of matsutake!"), Inoue's recordings yield moments where people explain that they do not have the proper body to discern the subtleties of matsutake smell. This negativity is itself a poiesis, of course, another synthesis of self and sense—here, something like a committed insensibility fomented in tandem with an elusive aroma that marks an unrefined sensorium as an apparatus of attunement not tuned to matsutake worlds. Matsutake becomes a field of discernment, of 'distinction' in Bourdieu's (1984) sense, where discernment of smell distinguishes people in the know.[1]

In this social economy of sense and self, the matsutake itself becomes rather vaporized. I say this not to suggest that matsutake disappears, but rather to note that its ontology is done precisely as a thing whose most significant force is not tangible, but atmospheric, ephemeral, and hard to pin down. Not only are matsutake hard to find (as they are) or hard to cultivate (as they are), they are also hard to language and hard to smell. In its suspensions found and induced, matsutake manifests qualities of relative intangibility, relative insubstantiality, and relative lightness or dispersal.

I repeat 'relative' in that last sentence in order to emphasize what is hopefully an obvious double point. That point is, first, that atmospheric things (things that belong, pertain, or are proper to the air) have the quality of being hard to substantiate, to weigh, to find. They require and drive the development of new sensory disciplines and prostheses. Second, this quality of elusiveness is relative. More precisely, the judgment of atmospheric things' elusiveness— the very sensation of their being hard to substantiate—is made in relation to norms of assessment, registration, and existence that are tuned to solid objects of sufficient mass, density, and size, to register and persist in one location or duration. Atmospheric things reveal themselves at the encounter of air and apparatus, and the pursuit of atmospheric adequation involves the development of different apparatuses, habits, and bodies of/for attention.

So what happens when people try to pin down such an atmospheric quality? If we think of efforts to do so as a kind of chase after atmospheric elusiveness, then this chapter can be considered a chase after that chase. Crucial in this chapter will be attention to the kinds of specific apparatuses and sensory trainings through which becoming atmospheric can be a means rather than a limit to apprehension. Put another way, while this chapter participates in an anthropology of elusivity via atmospheric phenomena, its aim is not to pursue an explanation of the conditions, meanings, or components of atmospheric elusivity. Instead, it seeks to learn from moments and efforts where parties learn to conform with the particularities of volatile and suspensive processes in order to sense and know something new about the world. These are moments where 'atmospheric attention'—attention attuned to the airborne, to substance in suspension (Cattelino, n.d.; Choy and Zee 2015), and to how things become so—becomes a condition of knowing. 'Suspension' here does not name a lapse or an emptying, nor does it denote a deferral of what matters, where resolution is elusive. Rather, here suspension means the very particular and concrete processes by which things and substances pass through and can be induced through different phases. I am interested, for instance, in how a solid can become an aerosol and, in being so suspended, how it helps the elusive come into view, magnified or reduced for sensory apparatuses still in construction.

Such developments are developments of method. Questions of method are questions of relation—the relating practices that condition knowledge of an object, as well as the conditions conditioning a particular manner of relating. At stake in this chapter's effort is an anthropology of atmospheres wrought in relation to elusivity itself, the salient problem of things that are definitionally hard to grasp, like a smell that is still just beyond the ability of words to describe. I mention this not to argue for the primacy of relationality over (object) ontology, but because the question that most interests me here is not the actuality of atmospheric objects or worlds, but rather how people adjust relations and relational capacities when motivated by an atmospheric question.

There is a growing body of work in anthropology and science and technology studies on atmospheric sensation (Calvillo, n.d.; Murphy 2006; Shapiro 2015; Zee 2017). As airborne toxics and pollution have become increasingly stabilized as matters of political and scientific concern—a recent WHO (2014) study, for example, describes air pollution as the number four cause of death worldwide and "the world's largest single environmental health risk"[2]—these researchers have begun to turn from arguing for air's materiality or substantiation (Choy 2011) to focusing instead on the varied forms of attention through which an atmospheric surround comes to sense and sensibility. Nicholas Shapiro (2015), for instance, attends to subclinical corporeal chemical knowledge among people living in conditions of formaldehyde exposure, where bodily sensations can attain fine-grained capacities to index small differences in atmospheric concentrations. Meanwhile, a different relation between instrumentation and bodies is activated in Nerea Calvillo's (n.d.) data visualization project "In the Air," which renders flat data files as an interactive digital airscape map of Madrid, where meshes undulate to reflect not only changes of particulate density over time and space, but also differences between interpolation algorithms. Data visualization is thus a misnomer, for the purpose of "In the Air" is to make data and data work palpable. The proposal is that data installation might elicit in a viewer/user a different kind of body and attention. This is ludic corporeal work.[3]

Shapiro and Calvillo teach us to approach environmental or atmospheric sensing as a problem of 'atmospheric sensation', where a distinction between data and corporeal experience, and the usual privileging of the former over the latter, is held in abeyance. This characterization of atmospheric sensation is important for several reasons. First, research protocols and clinical analytics not only condition the possibility of visibility for health effects of certain atmospheric substances (*pace* Fortun 2001; Petryna 2003); they also produce "historically specific terrains of invisibility" (Murphy 2006: 111), norms whereby some forms of effect cannot be substantiated as significant, causal, or specific (see Fortun 2001; Murphy 2006; Petryna 2003). Second, characterizing environmental or atmospheric sensing as a problem of atmospheric sensation foregrounds how numbers and measures themselves are both visceral and affecting.[4] Finally, by attending to how people reconcile diffuse symptomologies of exposure or discrepant visualizations dependent on algorithms and sensor location, Shapiro and Calvillo point to an emergent body of practices and methods for orienting to the tendencies of materials held and circulated in suspension.

Methods for an Atmospheric Anthropology: Tending to Suspension

Along with new techniques for attending to atmospheric substance come new ways of conceptualizing attention and knowing. This chapter explicates some of the sensory apparatuses and conceptual work that emerge in moments of 'tending to suspension', via the atmospheric attunements of matsutake lovers. Matsutake worlds offer a valuable way into the formation of capacities for atmospheric sensation, for matsutake's elusive atmospheric qualities catalyze a number of self-conscious activities of human-atmosphere relating. I think here of fieldwork discussions with a chef in the mushroom aisle of a Japanese market about the meal he planned to cook as he held matsutake and truffles to my nose; group practice with members of the Matsutake Worlds Research Group (MWRG) in Santa Cruz comparing the smells of fresh matsutake, dried matsutake, and matsutake crackers; MWRG experiments in Oregon and Japan locating mushrooms by smelling the soil; and a few successful and failed experiments in making spore prints from mushrooms of different types.

These atmospheric engagements are obviously not as laden with the freight of body burdens, carbon counts, dust drifts, or the breathtaking stakes of an atmospheric uncommons that undergird the broader concern for atmospheres mentioned above. They orient us instead to pleasures in atmospheric attunement—another structure of desire in the making of apparatuses. And perhaps, in the relative uneventfulness of their content, these scenes can allow us to focus on the form of some of the processes and techniques that emerge for characterizing and acting upon atmospheric phenomena.

In this chapter, I consider moments of tending to suspension where research and discernment of smell and spore are made possible by 'apparatuses of attunement', arrangements drawing human and non-human bodies and capacities together in such a way as to condition the possibility of a sensation.[5] I explicate such processes by drawing on food chemistry and field biology publications. I then highlight these processes and their significance by reconstructing the experiments in speculative narrative. Speculative reconstruction based on careful readings of scientific literature is a method I first developed to write about botanists and shifts in the ecopolitical sciences of specificity and endangerment in Hong Kong (Choy 2011). The method does not provide access to the cultural or social dynamics of the laboratory and field spaces or to the histories of particular individuals, so I will not comment on those. It does, however, enable a revival of scientific literature toward a different end. When water is added to dried mushrooms, they become more fragrant, not identical to their original state, but familiar and in some ways more potent. Similarly, my particular aim in the reconstructions that follow is to reanimate the structures of the experimental apparatuses that are hinted at in lists of equipment, measurements, and procedures in the literature.[6] In rendering these scenes, I intentionally narrate the subjects for which atmospheric qualities become present—those whose orientations offer the horizon of something's sensibility—as interlacings of equipment, objects, and structured attention. In particular, I linger over and highlight certain processes for tending to atmospheric suspension that accompany atmospheric discernment—techniques and methods such as concentration and reduction. I refer to these methods for atmospheric attention and relation as 'tending to suspension'.

Tending to suspension is a characterization I adopt to hold together a number of bundled issues pertinent to atmospheric qualities and relations, and the chase thereof.[7] The first bundle comprises the relations, substances, and conditions of atmospheric suspension—how things hang in an atmospheric medium. With Jerry Zee, I discuss conditions and qualities of suspension in more detail elsewhere (see Choy and Zee 2015), so I will only briefly comment on a few things here. First, suspension denotes an atmospheric hanging—a holding of something as an object of concern, such as a particle like PM_{10}, a spore, or an aromatic volatile. Second, suspension necessarily connotes more than one substance; one is held in another. The first is denser than the second, but somehow held aloft, while the second is a medium for the first. In attending to suspension, then, neither the first nor second substance can be thought of as alone; it is precisely their co-presence in a volume that matters. And so, it is the properties of both substances that will come to matter too. Suspension implies not a vessel or a container, but a current.

As for the conceptual bundle of 'tending', the word has a dual sense that I seek to exploit. The first is a sense of care and cultivation,[8] in the sense of tending a garden. This is a tending that is an attending, a form of accompaniment and patience—a practice akin to what Zoe Todd (2017: 107) has characterized as "tending to the reciprocal relationality we hold with fish and other more-than-human beings." Tending to suspension thus suggests a sense of ethical practice or technique in atmospheric care. The word's second sense is one of inclination, as in 'tending toward'—a leaning and sense that, under specific conditions, things and materials may develop a tendency to move in a direction. A tendency is more and less than a subjectivity or even an agency in matter. It is instead a property of things that emerges in relation to ambient conditions. Perfumes on warm skin tend to suspension. Tending to suspension bundles the relations, substances, and conditions of atmospheric suspension, while also pointing to the emergent practices and arrangements prompted by atmospheric concerns.

First Suspension: Reductive Apprehension

In this section, I discuss some scientists' recent experimental efforts to characterize matsutake aroma-active compounds. The laboratory processes/practices presented in the paragraphs that follow are not extraordinary in organic chemistry, but by elaborating the details of technique and equipment, I aim to highlight and name certain procedures in food chemistry that allow the negative and the diffuse to leave a (specific) trace. These procedures, concerned with volatiles, do condensatory and temporal work. They reduce matsutake to aroma—literally. The process includes evaporation and reduction (imagine a cook reducing a sauce). Reduction—a physical process by which the solid mushroom is converted and concentrated into a suspension of volatiles—is here a way of accessing the mushroom more fully, a way of thematizing and not erasing the sensory complexity of the human-mushroom encounter.

In 2006, a team of South Korean flavor scientists published an article, "Characterization of Aroma-Active Compounds in Raw and Cooked Pine-Mushrooms (Tricholoma matsutake Sing.)," in the *Journal of Agricultural and Food Chemistry* (Cho et al. 2006). It was the first of several publications on matsutake smell. A short while later, they published another article differentiating the flavor and smell profiles of different grades of matsutake (Cho et al. 2007). Among other things, the researchers found in that second study that umami—the vaunted fifth flavor of deliciousness that supplements the usual four terms of flavor description (sour, sweet, salty, and bitter)—and its chemical correlates were discerned in higher concentrations in Grade 2 mushrooms than in Grade 1. Another article soon followed, this one looking for umami differentials across grades and matsutake parts.

Let us focus on the second study by Cho et al. (2007). It puts us in the middle of something interesting. At this point, the laboratory has characterized the aroma-active compounds. They are now trying to distinguish the profile of compounds, compared across grades of mushrooms. And they are trying to establish 'instrument-sensory correlations', specific correlations between what people smell when they smell matsutake and what a laboratory instrument might be able to discern.

The authors set the stage by discussing the importance of a foodstuff's aroma. We learn that relatively few volatiles actually activate the odorant receptors; these are the aroma-active compounds. Then, as we enter the experiment and the transformation of matsutake aroma into aroma-active compounds, the discussion develops a temporal dimension. The life of matsutake volatiles are put into time. Time will be employed—solicited—to measure matsutake's aroma. In the process, the temporality of smelling will itself be affected.

Take the Korean grading scale for mushrooms. Grade 1 is the highest quality, over 8 centimeters long with an unopened pileus. Grade 2 mushrooms are generally 6–8 centimeters long, but their widths are irregular, and their pilei are not opened. Grade 3 are less than 6 centimeters long or have one-third opened pilei. Grade 4 mushrooms have completely opened pilei.

Matsutake are categorized into four grades in South Korea, as they are in Japan, according to their desirability in the market. These grades refer to the aging of a particular mushroom, its life passage, as much as its physical form. The most valuable matsutake are the youngest, those least damaged by the vagaries of fungal life. The veils on the cap are completely closed. To qualify for the two highest grades, for instance, a mushroom must have an unopened pileus. While both of the first two grades have unopened caps, they are distinguished from each other by length and regularity. A Grade 1 matsutake is a majestically long and intact thing, while a Grade 2, although almost as large and equally whole, widens and narrows along its body. Whether straight or veering, Grades 1 and 2 bear no incidental scratches or gouges. More suggestively, the membrane covering the gap between a mushroom's growing cap and its shaft must be unbroken. If that thin layer of mushroom flesh is compromised, it drops immediately to Grade 3. The movement from Grade 3 to 4 reflects the passing of matsutake time, the gradual opening of the cap. These are the first cuts of matsutake in this particular olfaction project, temporalizing grading cuts. They will yield the grounds for later comparison.

Other cuts, other grounds, follow, using techniques publicized in Cho et al. (2006). The mushrooms, divided by grade, are frozen to -70°C for storage. (They had only used Grade 1 mushrooms for the 2006 study.) For the experiment, frozen mushrooms are thawed and sliced. The next step sounds delicious: some of the sliced mushrooms are broiled, "heat-treated at $190 \pm 3\,°C$ for 1 min on both sides in a convection broiler (Toastmaster, Boonville, MO)" (ibid.: 6332–6333). The researchers now freeze both the raw and the cooked mushroom slices in liquid nitrogen. After this, they grind the frozen slices in a blender to attain a fine matsutake powder. These mushroom grounds are then swirled in dichloromethane to make an infusion, a mushroom tea, which is then filtered. The solids, the grounds, are left behind. The cutting, pulverizing, and infusion are key moments of a technical practice of attention and conditioning of material conditions for a substance to go atmospheric. These are material techniques for tending to suspension.

Let us return to the mushroom extract. The aromatic concentration is now subjected to an atmospheric engagement and transformation—a vacuum, which draws out the air above the tea and, crucially, what that air comprises. That is, it claims the volatiles, whatever will vaporize from the matsutake extract. The vacuum pulls fragrance from the extract that was itself pulled from ground slices of the solidity of the mushroom body. This matsutake air is now dehydrated and concentrated still more, yielding an even richer extract. This extract is vaporized again in a gas chromatograph, injected into a carefully heated chamber through which a stream of helium is running at a constant velocity. The sample vaporizes, and mushrooms, processed from bodies into extract, are again coaxed into particles in a gaseous medium. The sample becomes dispersed and suspended in a streaming mixture.

Apparatuses of atmospheric attunement require and yield a tending to suspension, an attention both to substances in suspensive relation and to the conditions under which they achieve such relation. Matsutake in the smell chemistry laboratory have elicited olfactory attention, and this attention has in turn rendered matsutake more volatile in order to smell it with a new smelling apparatus. Put simply, laboratory practices and apparatus tend to and hone matsutake's atmospheric tendency. The apparatus assists attunement or is attuned to olfactory attention, but it also reworks the mushroom in order to render it sensible.

The gas chromatograph through which the concentrated matsutake aroma streams is an apparatus that is a staple of chemistry research. "It separates perfume mixtures into their component parts," explains fragrance physicist Luca Turin (2006: 39). Or, perhaps more poetically, "this machine is to smell what a prism is to light" (ibid.). The gas chromatograph is a spatial and temporal technology—or, to put it more accurately, a spatializing and temporalizing technology. The chromatograph transforms vapor (a collection of substances in suspension, usually apprehended simultaneously) into a string of discrete substances, given different velocities across a narrow space, arriving at a desired destination—a sensor—at different times.

How does it work? The vaporous mixture of mushroom extract and helium is directed through a capillary—a long, narrow tube with thin walls—whose interior is coated with a silicon-based wax. At the other end is a detector (or detectors, as we will see), waiting for whatever comes out. The wax is 'sticky'; it slows the gas down, but differentially. Different compounds composing the gas get stuck differently along the way. Sticking, releasing—they all slow down, but some more than others, depending on their structure. Some go all the way through the capillary before the others.

So what results, the chemist hopes as she tinkers with the temperature, the composition of the mobile phase (helium gas) and that of the stationary phase (the coating on the tubes), is a drawing out of an aroma's time through a spreading out in space—a detection of different things in series, arrivals punctuated by pauses. The pause between signals makes the difference. The elapsing of time before a significant detection enacts the reality of distinctions, between signals, of a series of discrete substance(s).

We are getting close. I feel antsy because it has taken some time to get us here. But then again, this sensation of things dragging out—or, better, the 'dragging out of sensation'—is the crux of what is happening. An undifferentiated duration, a whiff, a lingering mingling simultaneity of smells is becoming a string of consecutive events through this carefully assembled tending to suspension. It is the smells' succession, their seriality, that will make them accessible to the equipment waiting at the other end of the tube.

What awaits at the end of the tube? There is a split in it. On one side is an FID, a flame ionization detector. It will flare and change in color as segments of air pass by. I am intrigued but do not yet know the history of how scientists calibrated the changes of the flame to particular compounds. This is still a black box to me—and accepted as such by the chemists. The flame is now enclosed, hidden from human view, while detectors monitor the color changes of the flame. At the end of the other tube stands a technician. She waits with her nose at the cone of the Olfactory Detector Outlet II (ODO II) (see fig. 3.1). With seven comrades, she has gone through a consensus training to agree on 15 olfactory descriptors for matsutake (see table 3.1), smelling unvaporized mushrooms across the four grades, discussing their smells, and agreeing on exemplary samples. She has also trained for over 30 hours with the ODO II. She waits at the cone as air streams out. Imagine her there.

Does she smell anything? … No.

Now? No.

Now? Yes! It is herbaceous! If all goes well, right at that moment the needle on the mass spectrometer at the end of the other line has also jumped. Something has landed.

Now? No. Wait.

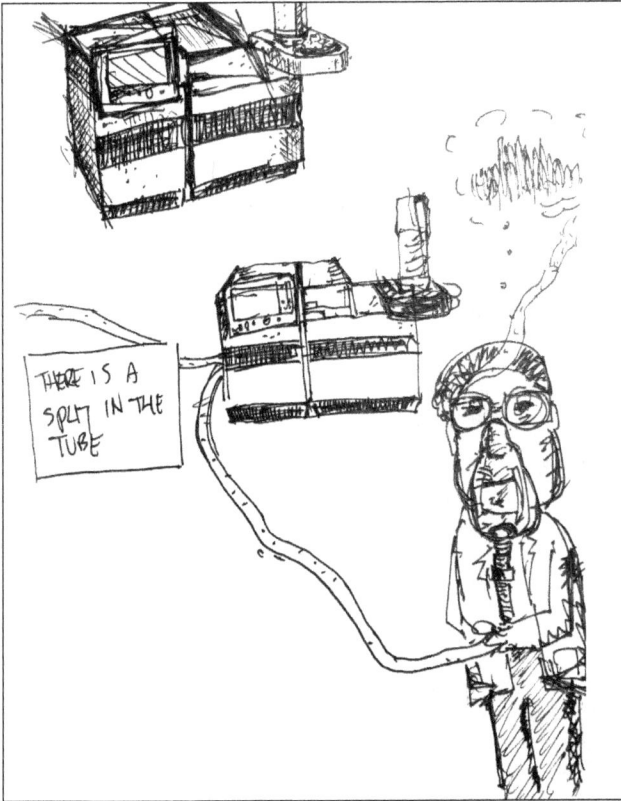

FIGURE 3.1: The split at the end of the tube (drawing by the author)

Now? No ... Now? Yes! It is floral, sweet! And then nothing again. Then, now! Something smells like cut grass!

Slowly, slowly, the vaporized mushroom splits into recognition as pieces. An aromatically elusive atmosphere, a complex simultaneity, becomes isolated smells in series.

What can we learn from this scene? There is first an obvious point that the apparatus is a sensory prosthesis. Scientific measurement here offers an extension of human sensorium, one that channels and disciplines what can become sensible in a particular way. But also, as Rachel Prentice (2013) and Natasha Myers (2015) have argued, living bodies serve and become animated as lively instruments in such experimental arrangements. The apparatus of atmospheric attunement makes new things and distinctions sensible—and that apparatus cultivates a sensing body, a sniffer. It not only assembles sensing bodies, but conditions kinds of sensibility.

TABLE 3.1: Definitions and references of the descriptive attributes for pine-mushrooms

Attribute	Definition	Reference
sweet	fundamental taste elicited by sugars	sucrose (1.5% in water)
salty	fundamental taste elicited by salts	sodium chloride (0.4% in water)
sour	fundamental taste elicited by acids	citric acid (0.05% in water)
bitter	fundamental taste elicited by caffeine	caffeine (0.07% in water)
umami	fundamental meaty taste elicited by monosodium glutamate (MSG)	monosodium glutamate (0.15% in water)
piny	aroma associated with pine needle	pine needle tea (one tea bag infused in 100 mL of boiling water for 2 min.)
floral	aroma associated with chrysanthemum	chrysanthemum tea (one tea bag infused in 200 mL of boiling water for 2 min.)
alcohol	aroma associated with rum	white rum (0.125% in water)
meaty	aroma associated with cooked meat	boiled beef
moldy	aroma associated with typical mushroom	1-octen-3-one (1000 ppm in dichloromethane)
wet soil-like	aroma associated with damp soil	damp soil
fishy	aroma associated with fermented anchovy	fermented anchovy sauce (0.025% in water)
fermented	aroma associated with Sikhye (a traditional Korean fermented rice beverage)	Sikhye drink (50% in water)
metallic	aroma associated with metals	stainless steel spoon
astringent	mouth feeling associated with tannins	tannic acid (0.1% in water)

Note: Reproduction of a table from Cho et al. (2007: 2324).

This insight recasts a moment early in our MWRG collaboration, when, like pupils who learned to be 'noses' for the French fragrance industry through *malettes à odeurs* (odor kits) (Latour 2004; Teil 1998), members of MWRG gathered to learn with each other to distinguish what makes matsutake smell like matsutake. Many of us were, like Inoue's humble witnesses, initially unable to smell anything distinguishing. But aided by a number of things—the experiences of those who knew the mushroom well, inherited phrases like Arora's 'red hots' and 'dirty socks', words we conjured and debated on the spot, and the table of cut mushrooms and artificially flavored mushroom snacks—we gradually learned to smell matsutake. The mushroom pieces, the crackers, the wrappers, the gathering, the plates, the knives, the palate-clearing sips of water—these, like the laboratory apparatus I have described, comprised an apparatus of attunement, an experimental setup entraining us to become researchers with particular sensitivities.

In addition to the point that the apparatus extends and elicits a particular sensory capacity, there is the matter of reduction and concentration. By this, I mean to call out from this arrangement the ways that reduction and concentration emerge through methodical practices of grinding and evaporating, material practices that tend to mushroom bodies as the starting format of an eventual suspension. These orient us to abstraction as a concrete practice of intensification.

Rather than an abstraction imposed from outside, abstraction is here elicited as a compounding of certain qualities. I am reminded of the diagrams that Elaine Gan and Anna Tsing think with in this book; their layered landscape tracings show that simplifications do not do simple work. Like them, I approach abstraction as a material practice, an intensification of a contour, rather than a taking leave of the world. Here, in the work of distilling arrangement, reduction is always rooted in technique, a mode for sensing and knowing by concentrating a quality or effect. Chemical characterization as a practice of olfactory discernment does a violence, breaking and grinding the mushroom body. Yet at the same time, it is an amplifying fidelity.

These material practices of reduction condition what we might call a 'reductive apprehension', by which I mean both the capacity for something to present itself more intensely as an object of experience or sensation and the apprehensiveness one might have around the transformations or leavings necessary for a particular form of concentration. Thinking through reductive apprehensions and the particular apparatuses of attunement they animate offers a conceptual workaround to the oppositional exchange between misgivings about 'reductionism' and declarations of urgencies to simplify, and perhaps, as well, a method for grounding theorizations of sensation in particular practices, apparatuses, and senses.

Second Suspension: The Chase and the Fail

If the loss of the solid mushroom is a condition for atmospheric knowing in the smell chemistry laboratory, it is a condition for laughter in my second scene of matsutake chasing. In a short YouTube video posted by a matsutake hunter, a dog is being trained to sniff out matsutake. The clip begins with a man's disembodied voice addressing a brown dog wearing a backpack. The camera puts us, the viewers, up high, looking down at the dog's dark brown face. The dog turns away and trots ahead through forest undergrowth. We follow. Without notice, the dog dashes off-screen; moments later, it doubles back to the left. The camera whips after it, and soon we find the dog waiting for us, panting proudly, sitting next to a round white shape half-buried in the dirt. "What?" the cameraman chuckles. "Good girl!" we hear. "Good girl! Don't step on it." We next hear rustling, and our point of view gets bumpy—the camera holder is getting ready to harvest the mushroom. "Sit!" he commands. "You found matsutake." Then he mutters, "Yeah, I *wish* you could smell them … too stupid." This utterance, along with the fact that the mushroom cap was already showing, are our main clues that this is a training run rather than a live feed of a mushroom hunt. Then, unexpectedly, excitement peaks; the dog jumps up, and—as the camera holder cries "No, no, no!"—she lands a paw in the middle of the prize. "Oh!! Oh no! Get back! Oh, you broke it, oh, that'll teach me," cries the voice. The mushroom cap has shattered. The dog's tail wags (Cetuspa 2010).

The video has a great title: "Dog Finds Matsutake Pine Mushroom Then Sits on It—Fail." It has been viewed a respectable number of times—close to 4,000 the last time I checked—but that is nothing compared to other videos that the same user has posted: over 63,200 views for "Matsutake Pine Mushroom How To Find Tips Tricks Tools," and almost 117,000 views for "How to Find Matsutake (Pine Mushrooms) Vancouver Island." Viewers are apparently not as interested in the 'fail' video.

But I like it. It is so appealing—and funny—because it taps the excitement of the hunt and the chase, the promise of a new way of hunting. In the camaraderie of hunter and dog, we recognize immediately how hard these mushrooms are to find. As my friends in this book so wonderfully show, matsutake are nothing if not elusive: their appearance can be unpredictable, they defy cultivation, and their abundance is subject to all manners of contingency. While this dog-matsutake-tracker idea is not exactly genius, it is totally understandable and kind of exciting. Other matsutake lovers must have wondered the same thing. Like truffles, matsutake are famed for their scent. For the man and his dog, recovering matsutake is a matter of attunement to its atmospheric traces, a multispecies effort to coordinate attractions and sensitivities.[9] If pigs have successfully helped humans find truffles by following their scent, can some other non-human animal friends be enticed to help humans find matsutake?

For this thematizing of matsutake smell, the thrill of the chase and the good-natured public display of failure make for an entertaining couple of minutes of mushroom comedy. It is also captivating because we can imagine all the off-camera work—the training, the trials, the errors. Time. The work of attention, attuning, tending to smell, to the trail, to the relationship.

I am taken by this chase and attunement, the subjection of bodies to the chase, to the line, following the fold, the shadow, the scent. The dog and its owner might figure it out, eventually. Maybe they already have. But that interests me less than the fact that they have spent so much time and effort attuning to each other, to scent, to manners of tracing, tracking, and communing. The mushroom's elusive presence—still a sensory challenge to the non-connoisseur human and his untrained dog—animates human, dog, and their still inadequate attunements into the building blocks for mushroom-sensing apparatuses, all for the thrill and the tremendous material promise of the chase. These YouTube videos are so compelling to viewers who know this mushroom because a primary aspect of the mushroom's allure, as Lieba Faier (this volume) argues, is the unpredictability of its appearance. Subject to contingencies of weather, soil, and good luck, they are nearly impossible to cultivate. Location and timing are everything—it matters where the spores fall.

Spores—their massive multiplication and their ability to traverse great distance—hold promise for matsutake lovers. The spores are a powder, part of a natural asexual reproductive system for mushrooms whose medium of transport is the air. Usually, for someone who would domesticate a mushroom, this means one needs to interrupt the airborne moment. To inoculate a sawdust medium, for instance, as one does with shitake, spores are mixed into liquid to be rendered transportable. They are thereby suspended not in air but in liquid, and contained by the density of that new medium.

But matsutake does not flourish in such attempts, and this is part of its wild allure. Its reproduction, its multiplication, will not be diverted or directed. Instead, one needs to follow its appearances, something that is particularly hard to track or predict because its spores are subject to drift. Elusiveness characterizes matsutake in part because those tiny, necessary aspects of the mushroom, which affect when and where one might find another, drift from apprehension with the slightest breeze.[10] The human-dog sniffing apparatus, an apparatus aspiring to attune to matsutake volatiles, comes to sniff for clues in the windy gap where spores fly and reproduction control fails.

Third Suspension: Chasing Spores

"Spores are hard to pin down; that is their grace," writes Tsing (2014: 227). What happens when you try? I turn now to dwell on some of the activity associated with this effort. In this last section, I reconstruct an experiment conducted by foresters to reflect on the condition of capturing and counting the elusivity of airborne spores. The experiment, whose results were published in *Mycobiology*, the journal of the Korean Society of Mycology, aimed to capture matsutake spores' airborne movement and pattern of dispersal (see Park and Ka 2010). As in the first section, I speculatively narrate the experimental scene hinted at by details in the scientists' written discussion of methods and equipment in order to discern and amplify the work of tending to suspension.

How do you catch a spore? I imagine the researchers bent at the waist, hands braced on knees, eyes on the ground. Hunkered down like that, in broad-brimmed hats and sturdy shoes, they would have gently brushed some dirt and duff to the side to see if the flash of tan on the gentle slope was what they hoped for: a mushroom cap breaking ever so slightly through the blanket of green, red, and brown pine. There might have been other mushrooms around, but this matsutake was special, or quickly became so. The researchers next fanned out, scanning the ground, and picked the area clean of other mushrooms. I wonder where they went, those cleared mushrooms—what Toastmasters, solvents, soups, or dinners did they end up in? The remaining matsutake, a soloist now, becomes the center of activity.

The humans form a circle around it, while taking care not to touch it. Instead, they take off their backpacks, pulling out stakes, mallets, a tape measure. They hammer 16 stakes into the soil. Four go in a line heading downhill from the mushroom, four head uphill; then two more lines of four are hammered along the slope, in the two perpendicular directions. The mushroom is in the crosshairs. And there is more. The one-meter length of each post is interrupted by three platforms: one at ground level, one at the top, and one right in the middle. Sixteen stakes, 48 platforms. Then nothing happens, until one day—the day the cap begins to turn up. This is the day they have been waiting for. At 10:00 AM, someone carefully places 48 glass slides onto each of the platforms. Each slide has been painted with a thin layer of glycerin. It is clear and tacky.

The stickiness makes the slides act like flypaper. Forty-eight little rectangles of sticky stuff, spaced in increments, waiting for messages from the mushroom. Of course, mushrooms do not talk, exactly—nothing so volitional as a speech act. But like a shout or a whisper, it is an airborne transmission, subject to the winds. Somebody's job is to count the matsutake spores on each slide. The counts in each direction, distance, and height are compared to glean the extent to which the dispersal of matsutake spores depends on factors such as wind and slope. And it depends a great deal, it turns out. More spores go farther in downhill and downwind directions.

I found this a bit of a letdown, as far as findings go. It seems obvious that wind and gravity would affect the dispersal of spores. More interesting to me are some of the numbers offered along the way. Ninety-five percent of that isolated matsutake's spores were caught by the experimental system. Eighty percent of those spores were caught by the slides on the lowest levels. Matsutake spores do not go very far, it turns out. Most of them apparently tumble more than they take off, despite the accounts we sometimes hear of some lucky high-flying globe-trotters.[11]

The percentages are deceptive. The researchers seem to know this, for in their text they take pains to explain that while the apparatus has captured 95 percent of the mushroom's spores, some 50 million spores have gone into suspension, eluding the experiment (Park and Ka 2010). How did they come up with figure, this conjuring of 50 million spores on the loose? How did they know how many there were? They have counted the number of spores that landed on each of their 48 spore traps. But how do they guess the number that got away? Speculation is involved, certainly, to offer a count of things that could not be counted, those spores that eluded capturing measures. Still, it is a good guess, a grounded speculation, for it is based on having counted 5 billion spores from a comparable individual mushroom shortly before this forest trip. These 5 billion interest me a great deal. What transpires in an effort to count—let alone capture—a multitude of tiny things that tend to tumble away or take off in the wind?

Suspensive Solution

To address what counting 5 billion spores entails in practice, I describe a background experiment conducted by foresters. This background experiment required a different kind of suspension: researchers hung a matsutake above a piece of foil and then waited. They waited four days for whatever spores it released to fall to the foil. The air in this experiment must have been kept so still. The foil must have been very large. After four days, the spores on the foil were brushed, pushed, nudged, and funneled into a solution. No longer spores about to be airborne, they were now particles in a liquid suspension. From one suspension to another.

This is a reduction—a reduction of this matsutake's spore print—in the sense that it is a concentration. The spores were corralled from potential take-off into one form of (airborne) suspension and put into another—a liquid. In other words, they were kept from a state of inconsistent and turbulent dispersal across a large and windy volume, guided instead into narrow dispersal in a small container.

This act of concentration points to the significance of tending to a medium. Mediums matter, for suspension is a relation enacted between what is being held and what is doing the holding. In moving spores out of their airy medium, which enables their flight and dispersal, this moment of concentration does not simply bring spores closer together. It also employs the tensions of a new medium to hold them closer. If gas chromatography dragged aromatic molecules out, here the spores are gathered in to preclude their spreading across space. These acts of spreading and gathering orient us to a key feature of atmospheric attunement, namely, an occupation with what conditions enable substances to tend to more or less dispersed states. Concentration is a pragmatic workaround for what Shapiro (2015) characterizes as atmospheric non-eventfulness—the impossibility of pinpointing a determinate temporal and spatial object with atmospheric phenomena.

Of course, there was still the matter of counting. It is a menial task, as many laboratory tasks are. Its apparatus involves a grid, a very small one, etched into a piece of equipment called a hemocytometer. Hemocytometers are tools for counting things—first developed for counting blood cells. A key point to note is that hemocytometers do not count on their own; instead, like gas chromato-graphs, they discipline human technique in the laboratory. A hemocytometer asks researchers to subject a substance to a particular kind of transformation so that this substance will be amenable to measurement. Crucially for our purposes, it asks for a specific way to count large numbers of very small things, a practice of counting by re-suspending.

Viewed from above, a hemocytometer looks like a thick microscope slide with a grid drawn on it. Looking from the side, though, you will see a small space, a hollow under the cover glass. This hollow was carefully made to hold a precise volume—a small volume (10e–4 milliliters) of liquid—whatever you are counting in a liquid suspension. As I mentioned above, all of this matsutake's spores were brushed into a container of liquid to make a solution of known volume. Now a portion of this solution is pipetted out, squeezed into the cavity of the hemocytometer. And someone counts the spores held in this micro-suspension.[12]

My aim in dwelling on this background experiment and its method is to underscore the significance of techniques for tweaking and attending to mediums of suspension and degrees of distribution and concentration. To count the entirety of even this small portion of the mushroom's spores would take too long. Instead, one counts the spores in selected squares within the larger grid. Those counts are compared, and one hopes they will be similar. Similar counts suggest that the experimenters have achieved an even and representative distribution of the tiny things—an evenly stirred suspension—and that the count can be scaled up proportionally to arrive at a count for the mushroom's entire spore population.

This is a tending to suspension, a method that first requires suspending spores' suspension in the air by putting them in solution. This tending entails practices and forms of concentration and reduction, as well as an effort toward even distribution—disciplined agitating so that the suspension of spores might achieve a consistency. Comparing these techniques for apprehending spores to those for vaporizing, concentrating, and streaming/splitting matsutake smell, we see how suspension, a state of elusiveness in the field, comes to be adopted and adapted as a means to measure. Tending to suspension means attending to and modifying various suspensive conditions. The conditions of ease with which particles of concern move or hang in their medium and their distribution in that medium are both the problem and the answer to this atmospheric elusiveness.

Conclusion: Tending to Suspension

To conclude, what might be gleaned about atmospheric attunement or the chase of atmospheres from these scenes? These activities could be glossed as moments of capture. One could take these efforts to measure, to quantify, to predict, to reproduce, to characterize, and to specify as projects of pinning down what is elusive about matsutake. But capture of the elusive, the ephemeral, the atmosphere (or the attempt thereof) is not exactly what is happening here, just as it is not quite capture that we see in the dog clip that I discussed earlier. I do not mean that these scientific efforts will fail like that happy dog, ultimately crushing the object of its owner's desire even if they corner the mushroom. That kind of argument has it place, but it is not the point I wish to make here.

Instead, I mean to emphasize that, successful or not, these scientific captures will capture more than the matsutake—they are mutual captures, involutionary pulls.[13] The humans and apparatuses in these atmospherically inclined mushroom experiments attune to—and become subject to—the qualities of relation between mushroom and air. Capturing the researchers in these scenes, even as they chase after matsutake, reveals properties of suspension. The dog, the sniffers, and the spore counters have been made to pause, to attend to an airy medium and how things are carried in it. The capacity for such atmospheric attention comes as an achievement through painstaking arrangements and tunings of bodies and qualities, including noses, containers, conduits, temperatures, degrees of agitation, densities, volumes.

These are methods for tending to suspension, techniques to notice, elicit, and follow suspended states. Tending to matsutake in suspension is done in small acts or moments, a sniff and glance between dog and mushroom hunter; acts of grinding, steeping, and distilling; a change of tube coating or the modulation of temperature or speed of a carrying gas; the collecting, swirling, and counting of spores. Tending to suspension names a form of attention attuned to such small techniques and acts of care as well as the contingencies of when something becomes less or more prone to settle. This is a particular kind of attunement compelled by the significance of the dispersed and very small, the particle, the fleeting—as well as the conditions under which such minutiae hang together, spread, or condense in a medium, in a volume, in a place, in a body.

Rather than focusing on the fleetingness or dispersal of matsutake's atmospheric moments or proclaiming their end, I have highlighted in these speculative reconstructions the practices and arrangements through which people grapple with the qualities of things. Put another way, rather than being interested in atmospheric elusiveness for its own sake—elevating 'the atmospheric' as an antidote or outside to the solid—I have explicated some of the methods and apparatuses motivated by atmospheric forms of matsutake presence.

My aim in tending in this manner to these tendings has been neither to suggest that the technosciences develop a more specific, concrete knowledge of atmospheric things, nor to denounce the reductionism of their desire to specify or quantify. I have instead been more interested in working through the concreteness of reduction and counting as practices. Sitting with, and slowing down, the processes of reduction—such as cooking down a fragrant infusion to increase its concentration, waiting for spores to land, or stirring collected spores into small distributed volumes—reveals abstraction as something other than an antonym to the concrete. Instead, we encounter the thickness of concrete practices of abstraction. In these scenes, abstracting practices not only enact a certain translational work of elusiveness into specification. They also generate new temporal and spatial conditions for attunement, expansions and contractions of the time and space between smells and spores, throwing subjects and their sensorial attention into the middle of volumes and durations where dispersed particles move slowly in suspension—closely enough to encompass, far apart enough to distinguish.

Apparatuses of atmospheric attunement provide some cues for methods of studying diffuse or suspended phenomena. Different experiments in rearranging concentrations and distributions of a thing enable different kinds of reductive apprehension, as do adjustments of the mediums that carry it. Anthropologists of elusiveness might study the apparatuses of attunement that elusive objects elicit, or perhaps develop apparatuses of our own, with special attention to the material and conceptual coordinations by which we generate our own apprehensions. These reductive apprehensions generate and rely upon situated practices of abstraction that momentarily heighten particular qualities. These are practices of abstraction that work by following in proximity rather than viewing from a distance; they yield not simply the temporary stilling of an elusive object in space and time, but a simultaneous subjection of the student to the material and conceptual terms through which an object's elusivity is felt as such.

In emphasizing the generativity of atmospheric experiments for fostering new kinds of sensation whose effects exceed a project of explanation or exegesis, I echo and bend arguments made by interlocutors, including Brian Goldstone, Stuart McLean, Anand Pandian, and Robert Desjarlais, at a recent "Image as Method" symposium (Romero 2015). In this exchange, elusiveness figures centrally, both as a descriptor of phenomena that might be construed as non-empirical and as a characteristic of images that an 'imagistic' anthropology might put toward rendering such phenomena. Elusiveness is presented as a quality of liminal[14] or fantastical[15] realities that go beyond conventional discursive turns to evidence or categories. An 'imagistic' anthropology, then— by which Goldstone means an "elusive, often ephemeral and difficult" method of working through imagistic materials, where 'image' is not confined to the visual but includes, "for instance, dreams, sounds, particular linguistic forms, modes of writing, etc." (cited in Romero 2015)—thus becomes an appropriate tool for approaching and rendering such hard-to-grasp qualities and states. My method has been to approach atmospheric elusiveness with a similar sensibility—one that would pause before explaining elusiveness away, yet would do so within the scenes, terms, and practices of empirical tracking projects that initially appear engineered to reduce ambiguity and diffusion. Such projects, along with the apparatuses of attunement geared for them, pragmatically embroil a subject into atmospheric tendencies and tendings. Rather than suspending analysis, they invite one to ask how one might learn and dwell in the terms and conditions of particular suspensive moments. In atmospheric tendings, the notions of the empirical and the elusive convey each other.

Such tendings open ways of 'being-in-world', among smallnesses that make a difference, in mediums that hold together and apart. Not 'being-in-the-world'—no definite article, no defined world—for the moments amplified are ones of leaving an already solidified world to expand an indefinite space between signals. The pursuit of atmospheric elusiveness here yields not just an attainment of the object of the hunt, but a stirring of the chaser into new relations and techniques of attention. These are scenes of subjects and substances tending to suspension.

Acknowledgments

Warm appreciation to Lieba Faier, Elaine Gan, Michael Hathaway, Miyako Inoue, Shiho Satsuka, and Anna Tsing for the years of exchange that underlie the writing of this chapter. I am also grateful to audiences at the New School, Oxford, Chicago, the Chemical Heritage Foundation, Texas, Rice, and Stanford who heard and critiqued earlier versions. For intellectual kindnesses that enabled this chapter's formation, I extend thanks to Amy Hinterberger, Anne Walsh, Chris Kelty, Cori Hayden, Dehlia Hannah, Dimitris Papadopolous, Engineered Worlds, Hannah Landecker, Heather Paxson, Hugh Raffles, Jake Kosek, Javier Lezaun, Joseph Dumit, Joseph Masco, Kaushik Sunder Rajan, Marisol de la Cadena, Mei Zhan, Michelle Murphy, María Puig de la Bellacasa, Natalie Porter, Natasha Myers, Nerea Calvillo, Nicholas D'Avella, Nicholas Shapiro, Stacey Langwick, Stefan Helmreich, Stefanie Graeter, Suzana Sawyer, Vivian Choi, and Zamira Ha. Deep thanks to Lieba Faier, Joshua Weiss, and Jerry Zee for especially detailed readings and feedback that I found invaluable, as well as to Martin Holbraad for his principled engagement and pressings for explication that helped me to strengthen the chapter's arguments and structure.

Timothy Choy is an Associate Professor of Science and Technology Studies and an Associate Professor of Anthropology at the University of California, Davis. He is the author of *Ecologies of Comparison: An Ethnography of Endangerment in Hong Kong* (2011), which won the Rachel Carson Prize from the Society for Social Studies of Science. His interests include ecological discourses and practices, human-nonhuman relations, emergent politics of air and breath, concept methodology, and questions of form.

Notes

1. For a good elaboration and illustration of Bourdieu's approach to aesthetic disposition as a structured and structuring sensibility that distinguishes (between) subjects at the same time as it distinguishes (between) objects, see Bourdieu (1984). For more on Inoue's research and that of the Matsutake Worlds Research Group, see MWRG (2009a, 2009b).

2. Mainstream news sources have also chimed in, noting that "air pollution kills more than HIV and malaria combined ... With nearly 1.4 million deaths a year, China has the most air pollution fatalities, followed by India with 645,000 and Pakistan with 110,000" (Borenstein 2015).

3. On the technological cultivation of bodily orientations and attentions, see also Prentice (2013) and Myers and Dumit (2011).

4. See for instance, Joseph Dumit's (2012) work on the affective force of statistical 'thresholds' for health and illness.

5. Jerry Zee and I introduce atmospheric apparatuses as method in Choy and Zee (2015), as a methodological retooling of, among others, Barad (2007), Kortright (2013a, 2013b), and Rheinberger (1997).

6. Like the chapter by Elaine Gan and Anna Tsing in this book, this chapter could be understood as a grounded thought and sense machine. The method in both is to abstract and assemble scenes in order to throw into relief certain processes and qualities so that they can be open to reflection and conceptualization.

7. On the difference between writing to characterize or propose and writing to prove or unveil, see Stengers (2015: 33–34).

8. On care as a form of cultivated and cultivating practice, see Puig de la Bellacasa (2012). Studying efforts to test the viability of novel strains of rice in C_4-rich atmospheres, Kortright (2013b) highlights a particular form of care attuned to the conditioning and composition of atmosphere in experimentally controlled chambers.

9. In this book, Gan and Tsing propose diagrammatic approaches to marking the coordination of different ecological temporalities, while Hathaway highlights modes of multispecies attraction, including the olfactory attraction of flying squirrels and insects to truffles. Together, my collaborators teach me to recognize dog-human mushroom hunting as an effort to effect an intentional coordination of multispecies attractions and sensitivities. An apparatus of attunement assembles bodies and capacities as equipment.

10. For more on the elusiveness of matsutake reproduction and the generativity of thinking with the movements and genetic characteristics of matsutake spores, see Tsing (2014: 227–239).

11. See, for instance, Tsing's (2014) ventriloquism of a spore.

12. A count measures and beats. See, for instance, Kortright's (2013a) account of the activity of counting modified C_4 sorghum seeds, which narrates a rhythmic transition from concentration to a state of relaxed awareness: "I concentrated on each of the little seeds that sat in front of me. The seeds that went into the envelope needed to be good seeds. Goods seeds were those seeds that were

not seen as ugly. Ugly seeds were the ones that were broken in half, too small, or too shriveled. They were the ones that had a lesser chance of germinating. I looked at each of the seeds as I counted them, and I pulled out the ugly seeds. Sometimes I stopped and inspected the seeds. I would take them between my fingers and look at them closely, but usually I could see the ugly seeds as they [slid] across the paper. Scouring for ugly ones, I would hunch over really close to the seeds. I held the stick firmly. Everything the stick touched, I touched. I felt the touch of the stick on every individual seed. I was very precise with each seed: one ... flick ... two ... flick ... I would make sure every seed was counted ... three ... flick ... four ... flick ... I wanted to make sure every seed looked good, and I wanted to make sure they each would grow ... five ... flick ... six ... flick ... seven ... flick ... eight ... flick. After the first five envelopes, I started to relax. It was not that I stopped seeing the seeds; I saw them—the color, the texture. Like knitting, I saw with my fingers as much as my eyes, and although I was relaxed, I was still aware."

13. On 'involutionary momentum', see Hustak and Myers (2012), whose reading of Darwin's explorations of orchid anatomy highlights how experimental forms of attention and touch brought scientist and plant into novel forms of embrace.

14. On the paradox of studying 'liminality' via sociological or anthropological categories, McLean writes "it is society as an assumed structure of relationships and classifications that is prioritized as providing an explanatory framework for understanding the liminal, as we are led away from the elusiveness of the liminal itself and back to the (allegedly) more stable and knowable terrain of social relationships and cultural significations" (cited in Romero 2015).

15. Desjarlais says of the fantastic and phantasmagoric: "There's so much going on—there's so many layers of fantasy and the fantastical qualities of everyday life—and so I have been thinking about how one can write about that, because it presents a challenge anthropologically. If we are trained to attend to what's empirical and what one can provide evidence for, what about all these other modes of thought that are much more elusive and much more 'non-empirical' in a way?" (cited in Romero 2015).

References

Arora, David. 1986. *Mushrooms Demystified: A Comprehensive Guide to the Fleshy Fungi.* 2nd ed. Berkeley: Ten Speed Press.

Barad, Karen. 2007. *Meeting the Universe Halfway: Quantum Physics and the Entanglement of Matter and Meaning.* Durham, NC: Duke University Press.

Borenstein, Seth. 2015. "Study Ties Farming to Air Pollution Deaths." *Des Moines Register*, 16 September. https://www.desmoinesregister.com/story/money/agriculture/2015/09/16/study-ties-farming-air-pollution-deaths/32512361/.

Bourdieu, Pierre. 1984. *Distinction: A Social Critique of the Judgement of Taste.* Trans. Richard Nice. Cambridge, MA: Harvard University Press.

Calvillo, Nerea. n.d. "In the Air." http://www.intheair.es.

Cattelino, Jessica. n.d. "The Airborne Politics of Interdependency." Unpublished manuscript.

Cetuspa. 2010. "Dog Finds Matsutake Pine Mushroom Then Sits on It—Fail." YouTube. http://youtu.be/ZVnbMj9_RDk.

Cho, In Hee, Se Young Kim, Hyung-Kyoon Choi, and Young-Suk Kim. 2006. "Characterization of Aroma-Active Compounds in Raw and Cooked Pine-Mushrooms (Tricholoma matsutake Sing.)." *Journal of Agricultural and Food Chemistry* 54 (17): 6332–6335.

Cho, In Hee, Soh Min Lee, Se Young Kim, Hyung-Kyoon Choi, Kwang-Ok Kim, and Young-Suk Kim. 2007. "Differentiation of Aroma Characteristics of Pine-Mushrooms (Tricholoma matsutake Sing.) of Different Grades Using Gas Chromatography-Olfactometry and Sensory Analysis." *Journal of Agricultural and Food Chemistry* 55 (6): 2323–2328.

Choy, Timothy. 2011. *Ecologies of Comparison: An Ethnography of Endangerment in Hong Kong.* Durham, NC: Duke University Press.

Choy, Timothy, and Jerry Zee. 2015. "Condition—Suspension." *Cultural Anthropology* 30 (2): 210–223.

Dumit, Joseph. 2012. *Drugs for Life: How Pharmaceutical Companies Define Our Health.* Durham, NC: Duke University Press.

Fortun, Kim. 2001. *Advocacy after Bhopal: Environmentalism, Disaster, New Global Orders.* Chicago: University of Chicago Press.

Hustak, Carla, and Natasha Myers. 2012. "Involutionary Momentum: Affective Ecologies and the Sciences of Plant/Insect Encounters." *differences* 23 (3): 74–118.

Kortright, Chris. 2013a. "On Labor and Creative Transformations in the Experimental Fields of the Philippines." *East Asian Science, Technology and Society* 7 (4): 557–578.

Kortright, Chris. 2013b. "Producing Evolution: The Growth Chamber as a Novel Ecosystem." Paper presented at the annual meeting of the Canadian Anthropological Society, 10 May.

Latour, Bruno. 2004. "How to Talk about the Body? The Normative Dimension of Science Studies." *Body & Society* 10 (2–3): 205–229.

Murphy, Michelle. 2006. *Sick Building Syndrome and the Problem of Uncertainty: Environmental Politics, Technoscience, and Women Workers.* Durham, NC: Duke University Press.

MWRG (Matsutake Worlds Research Group). 2009a. "A New Form of Collaboration in Cultural Anthropology: Matsutake Worlds." *American Ethnologist* 36 (2): 380–403.

MWRG (Matsutake Worlds Research Group). 2009b. "Strong Collaboration as a Method for Multi-sited Ethnography: On Mycorrhizal Relations." In *Multi-sited Ethnography: Theory, Praxis and Locality in Contemporary Research*, ed. Mark-Anthony Falzon, 197–214. London: Ashgate.

Myers, Natasha. 2015. *Rendering Life Molecular: Models, Modelers, and Excitable Matter.* Durham, NC: Duke University Press.

Myers, Natasha, and Joseph Dumit. 2011. "Haptic Creativity and the Mid-embodiments of Experimental Life." In *A Companion to the Anthropology of the Body and Embodiment*, ed. Frances E. Mascia-Lees, 239–261. Hoboken, NJ: Wiley-Blackwell.

Park, Hyun, and Kang-Hyeon Ka. 2010. "Spore Dispersion of *Tricholoma matsutake* at a *Pinus densiflora* Stand in Korea." *Mycobiology* 38 (3): 203–205.

Petryna, Adriana. 2003. *Life Exposed: Biological Citizens after Chernobyl*. Princeton, NJ: Princeton University Press.

Prentice, Rachel. 2013. *Bodies in Formation: An Ethnography of Anatomy and Surgery Education*. Durham, NC: Duke University Press.

Puig de la Bellacasa, María. 2012. "'Nothing Comes without Its World': Thinking with Care." *Sociological Review* 60 (2): 197–216.

Rheinberger, Hans-Jörg. 1997. *Toward a History of Epistemic Things: Synthesizing Proteins in the Test Tube*. Stanford, CA: Stanford University Press.

Romero, Andrés. 2015. "Image as Method: Conversations on Anthropology through the Image." Somatosphere, 14 August. http://somatosphere.net/?p = 10644.

Shapiro, Nicholas. 2015. "Attuning to the Chemosphere: Domestic Formaldehyde, Bodily Reasoning, and the Chemical Sublime." *Cultural Anthropology* 30 (3): 368–393.

Stengers, Isabelle. 2015. *In Catastrophic Times: Resisting the Coming Barbarism*. Trans. Andrew Goffey. London: Open Humanities Press.

Teil, Geneviève. 1998. "Devenir expert aromaticien: Y a-t-il une place pour le goût dans les goûts alimentaires?" *Sociologie du Travail* 40 (4): 503–522.

Todd, Zoe. 2017. "Fish, Kin and Hope: Tending to Water Violations in *Amiskwaciwâskahikan* and Treaty Six Territory." *Afterall: A Journal of Art, Context and Enquiry* 43: 102–107.

Tsing, Anna L. 2014. "Strathern beyond the Human: Testimony of a Spore." *Theory, Culture & Society* 31 (2–3): 221–241.

Turin, Luca. 2006. *The Secret of Scent: Adventures in Perfume and the Science of Smell*. New York: Ecco.

Volk, Tom. 2000. "*Tricholoma magnivelare*, the American Matsutake Mushroom." University of Wisconsin Plant Teaching Collection. http://botit.botany.wisc.edu/toms_fungi/sep2000.html.

WHO (World Health Organization). 2014. "7 Million Premature Deaths Annually Linked to Air Pollution." https://www.who.int/mediacentre/news/releases/2014/air-pollution/en/.

Zee, Jerry C. 2017. "Holding Patterns: Sand and Political Time at China's Desert Shores." *Cultural Anthropology* 32 (2): 215–241.

Chapter 4

MATSUTAKE, SO AROMATIC IN ITS ABSENCE

Miyako Inoue

The Smell of Charisma: The Matsutake Mushroom

This chapter is about the nose. I want to discuss how language interacts with the nose's way of apprehending the world and brings forth olfactory experience. My focus is the smell of the matsutake mushroom (*Tricholoma magnivelare*), which is highly prized in Japan for its distinctive piney and earthy aroma. It is a mushroom with 'charisma' (Satsuka 2013; Tsing 2015). Philosopher Watsuji Testuro (1995: 70) recollects mushroom hunting as one of the most pleasurable activities of his childhood and, above all, the sublime experience of a rare encounter with matsutake, "lifting up dead leaves and rising from the ground, with the king's dignity and the saint's fragrance." For Watsuji as a child, finding matsutake was "as big an event as when a scientist discovered radium" (ibid.). The matsutake's olfactory quality stands for its culturally prominent way of being. What kind of smell is that of "the king's dignity and the saint's fragrance"? What is the smell of charisma?

Over the years, the Matsutake Worlds Research Group (MWRG),[1] of which I am part, has carried out fieldwork in Oregon, Canada, Japan, Korea, and Yunan, China, on the global matsutake commodity chain and its environmental,

Notes for this chapter begin on page 87.

political, and social impact upon the relationship between humans and matsutake (see, e.g., MWRG 2009). This project is informed by critical reflection on the human-centered approach that is less capable of effectively addressing environmental destruction caused by humans, and is committed to understanding the social worlds of the more-than-human, in which non-human species are to be recognized not as objects but as actors or agents. Accordingly, nature is no longer what the human represents through language: it is in and of itself, without language mediating and representing it to us. Let the nose speak; do not let the mouth speak for the nose.

For me as a linguistic anthropologist, this may be a self-defeating project. But it is also an opportunity to explore other semiotic modalities in which the human apprehends the social world. Sensory experience is precisely a case in point. How can something else—language—stand in for the sublime sensation of touching my cat's stomach other than that in and of itself? Nonetheless, I need language to know it and to represent it. So my position here is to remain in the social world experienced as and through linguistic representation, but to focus on semiotic forms and strategies in representing something that is not representable.

Matsutake grow around the roots of living red pine trees, and the local ecology that sustains the symbiotic relationship among matsutake, pine trees, humans, and other species is impossible to cultivate artificially. In Japan, matsutake are a seasonal delicacy of national importance whose first auction of the year predictably makes news, announcing to the nation the arrival of autumn and forecasting the given year's climate variation. The rapid decline in its production after World War II prompted the increasing presence of imported matsutake from elsewhere in East Asia as well as North America, Scandinavia, and North Africa. The scarcity of domestic matsutake has elevated not only their economic value but also their cultural importance as a distinctive national symbol.

Just as sensory qualities in general are linguistically conventionalized, the smell of matsutake is no exception. It is an excessively meaning-*ful* sign. Matsutake is a symbol of an olfactory trace of Japan's imaginary past of natural abundance and cultural plenitude. The smell of matsutake is memorialized by stubbornly banal language and the cultural associations it stipulates. One can repeatedly hear and read platitudes such as "The smell of matsutake is a sign of the arrival of the season of autumn," "Matsutake is the taste of Japan," "Matsutake is the smell of *furusato* [hometown]," or "Matsutake is the king of autumn taste [*mikaku no ōsama*]," and their variations. And if you are careful, you would not miss a small quotative attached to it, so that instead of "matsutake is a smell of *furusato*," you would actually hear, "It is said that matsutake is a smell of *furusato*," or variations such as "People say," or "We [I] often hear." Those clichés circulate and glide on a surface of everyday communication without anyone claiming their ownership. Or ask any Japanese adult about matsutake. I bet that he/she would never fail to say "Nioi matsutake aji shimeji," meaning that if you choose a mushroom for smell, pick matsutake, for taste, pick shimeji. It is a citation that does not perform any of the types of speech act, and citing it is a culturally conditioned reflex. Matsutake thus creates the 'Japanese nose'.

Most of my interlocutors were over the age of 70 and lived in the Kansai region of Japan, where matsutake had been abundant until the 1960s. I was hoping to listen to memories of the aroma of matsutake connected with personal histories, and, by doing so, to recapture indexical—singular and existential—meanings of the smell of matsutake that are not yet abstracted to the oversaturated symbolic sign of national habits. However, my expectation proved to be somewhat off. As the Japanese phrase goes, I was *noren ni udeoshi*, that is, 'punching through a cloth curtain', or beating the air. "Yes, it smells good." "Oh, I know that smell." Those were typical responses, and my respondents' narratives orbited around that which is not representable. I admit that my interview skills may certainly have been to blame. But I have also come to appreciate them as forms of poetics or aesthetic forms, which show us an alternative mode to represent sensory experience.

Smell: Qualia Caught between Language and Nose

The nose connects us with the external world in an unique way. Unlike some other 'interfaces' between the human body and the surrounding environment, such as the ears and eyes, smell enters and exits the body with no absolute boundary unless one does not breathe. Sound can be shielded by covering one's ears or shutting the window, and sights can be blocked by closing one's eyes. Smell mercilessly permeates. How long can you hold your breath when your nose is assaulted by an odd smell? But at the same time, smell is a fleeting event. It would be difficult to present a smell as evidence. Given these qualities of smell, the nose knows the world not as a collection of discrete objects but as a series of events with no distinction between humans and non-humans. To borrow from Shiho Satsuka's discussion in this volume of how humans develop their sense of being in the world through "multispecies entanglements" with non-human entities, the nose knows the world not as *mono* (things) but as *koto* (events). Accordingly, smell is a mere quality that is an undifferentiated part of an event.

If smells are parts of events, then the nose's contact with the world is ephemeral and atmospheric, with no discrete object to smell. Peirce (1978: 558) recognizes a sign in such a monadic state of being as a "qualisign," which has a conventional imputed quality or what Gal (2013: 33) calls "attributed quality."[2] In his discussion on how scientists try to develop methods and techniques to see and catch things that are not tangible, such as smell, Timothy Choy (this volume) proposes the concept of "apparatuses of attunement" as a variety of methods of experiencing the world, involving arrangements that draw "human and non-human bodies and capacities together in such a way as to condition the possibility of a sensation." It names an audacious attempt to know the world without the mediation of language (see also Zee 2020).

For non-scientists, however, it is here that language inevitably intervenes. The linguistic mediation of olfactory sensation thus separates that which belongs to the realm of *koto* into a subject and an object: the person who experiences a smell and the thing that is smelled. The monadic world of atmospheric attunement must be mediated and divided by language—a culturally specific set of words, categories, expectations, associations, and narratives—in order for the experience of smell to become parsable into a discrete object and its source. In language, smell loses its ability to stand for itself, and we lose our ability to access it directly.

Smell is thus always already a memory, an indexical memory of sense that allows one to be in the present through the past, or, very often, vice versa. I remember my first airplane trip, from Japan to the United States, by the smell of scorched coffee that permeated the aircraft. The Japanese verb *takishimeru*, which refers to an erstwhile custom of perfuming robes with burned incense, came to my mind. The smell of coffee emanated not only from the coffee pot but also from the walls and floors of the entire aircraft, as if it had been smoked in a coffee pot that had brewed coffee hundreds of thousands of times before. Since then, every time I step into a coffee-infused airplane, it immediately takes me back to my first trip to America. Something similar occurs with the smell of toothpaste and miso soup together. Although I do not often encounter this particular combination of smells, when it happens, it takes me back to the morning routine of my not-so-pleasant high school days. The olfactory sensation that travels through the nose thus becomes a robust indexical sign that connects now to then and here and there.

At the Tsukiji vegetable market, four of us from the MWRG group met a matsutake guru who is an intermediate wholesaler (*nakagainin*) participating in matsutake auctions and selling the mushrooms to high-end restaurants. He told us that he could tell the difference between domestic and imported matsutake, with the former prized far more than the latter. At the first auction of the season, the price of even a tiny mushroom is astronomical as long as it 'says' it is Japanese. How so? The *nakagainin* claims that the domestic matsutake has a clean piney smell, while the imported one has the smell of mud. Not all noses are equal. I tried, but my nose did not know the difference between Japanese and imported matsutake. Among the imported mushrooms, the *nakagainin* said that his trained nose could even tell the nationality of the matsutake, for example, the difference between matsutake from China or from

Korea or Morocco. The contrast between cleanliness and muddiness in this representation of the quality of matsutake is thus construed as an iconic contrast in the quality of national character between Japan and that of its Others. This is an exemplary semiotic process of rhematization (Gal 2005, 2013; see also Ball 2014). The matsutake guru talked about the imported matsutake—cheaper and, by his standard, lower in quality—as if these foreign countries were invading Japanese territory.

Matsutake, So Aromatic in Its Absence

Experiencing olfactory sensation is one thing, and talking about it is another. There is something intriguing about the way in which people talk about matsutake. Of course, I do not claim that everyone talks about matsutake in the same way, or that these are the only ways that matsutake can be talked about. But an interesting pattern emerges from my preliminary interviews. I had expected to hear more affective and sensual narratives. While all the people I interviewed claimed to love the smell of matsutake, their narratives have a noticeable and, given my research topic, frustrating absence of smell—a void that they circled around, as if matsutake, the charismatic non-human, demanded certain ethics and aesthetics of representation. It is as though the most authentic olfactory experience of matsutake required, paradoxically enough, its absence in representation.

Some of my interviewees remembered the annual community meeting where the rights to matsutake mountains was auctioned at the beginning of the season. These mountains were the community's common property, and the profit from the auction would be used for community purposes. The auction would start only hours after the participants got sufficiently drunk, hoping that the euphoria from intoxication would make buyers lose their calculating judgment regarding the bidding price and help fill the community's coffers. No one knows beforehand how much or even whether an auctioned mountain will yield matsutake in the coming season. In some years, it will abound while in others it will not. It is of no use memorizing the location of the trees where you found matsutake in any given year, as there is no guarantee that you will find them there the next year. Various theories circulate, yet in the end it is nothing but a wild and hopeful guess.

So what would you do with the matsutake you collected? Mr. Uchihashi told me that the mushrooms would be given to neighbors rather than consumed at home: "Matsutake is something to give to others. It is not a terribly tasty thing, anyway." He then added: "It is like money. It goes out and never stays!" For Mr. Uchihashi, what makes matsutake matsutake is that it must leave you. You might momentarily sniff its pleasant aroma as it passes by on the way to someone else. Unlike dried shiitake, a traditionally common *oseibo* (year-end gift) and *ochūgen* (mid-summer gift), only matsutake is culturally deserving enough to demand the care and cost to be exchanged fresh.

In old Japanese letter-writing manuals—which provide the reader with model letters for all kinds of possible occasions, such as wedding and funeral announcements, declining a matchmaking proposal, getting a job, moving out, and writing to brothers and fathers living in Japan's colonies—a letter for sending and receiving matsutake is an established genre. The model matsutake letter conveys the sender's mindfulness in sending the gift swiftly to keep intact its freshness and, most importantly, its aroma. For example, a model sender's letter would read: "These matsutake have just arrived from Kyoto today, and we wanted them to be taken to you as they are before they lose their aroma. Please grill them for your father's dinner" (Nakamura 1930: 43).

An example of a model thank-you letter for the matsutake gift would read: "It was only about three hours after the letter had arrived that the matsutake arrived. We were so happy that it had arrived so quickly. We opened the basket right away. You packed it so well that they were kept intact and so was their gorgeous fragrance, as if they had just come from the mountain" (Kawai 1912: 363–365). The letter writer goes on to describe her plan to share the mushrooms with her family's close neighbors, which she had kept secret from her brother, who instead had the idea to bring the matsutake to his friends to share with them. The letter ends with the ruse revealed and the family having settled on keeping the matsutake and grilling them for their own enjoyment.

The olfactory experience of a gift of matsutake was not limited to the mushrooms themselves. A *shida* fern is often placed under the matsutake in the gift basket or box, where its piney smell and vibrant green enhance both the matsutake's aroma and its earthy color. But in the old days—as another research participant, Mr. Fujiwara, told me—cash was sometimes hidden under the fern. That cash, he added, had to be in crisp, mint condition with no wrinkles or smudges. The cash dressed and smelled like matsutake, and the latter literally became the carrier of the (real) gift, not its content. Mr. Fujiwara made it clear that he himself had never received or sent a gift of cash-under-matsutake, but he reasoned that the amount of cash must be more than the monetary worth of the mushrooms. It is intoxicating to imagine the smell of bills fresh from the mint perfumed by matsutake and fern, or—what is equally sumptuous—matsutake infused with the smell of fresh bills.

An all-too-typical opening response to my matsutake questions began with the disclaimer, "It smells so good! It used to be around in the old days (but not anymore)." Or "In the old days we used to see a lot of them, but not anymore." Or "I used to eat it in the old days, but not now." Or "Yes, people say that it has a nice aroma ... that's not my direct experience, but someone else's." In a group of women in their late seventies whom I interviewed, none remembered whether she used to eat matsutake in her childhood in spite of the fact that all of the women grew up near matsutake mountains. One even confessed that she liked *matsutake no aji osuimono* (matsutake-flavored soup) better than real matsutake and that it was disappointing when she had the latter for the first time. As she recalled, she did not find it that "tasty." The others nodded and laughed, indicating that she was not the only one who had had this inverted experience. *Matsutake no aji osuimono* is the brand name of a synthetic soup powder made of 'matsutake extract'. It comes in an acutely recognizable package of an unnameable color (fig. 4.1).

After you pour boiling water into the *owan* (soup bowl) together with the soup powder, and if you look carefully, you can probably spot what looks like tiny bits of dried real matsutake. Having been around since 1964, by which

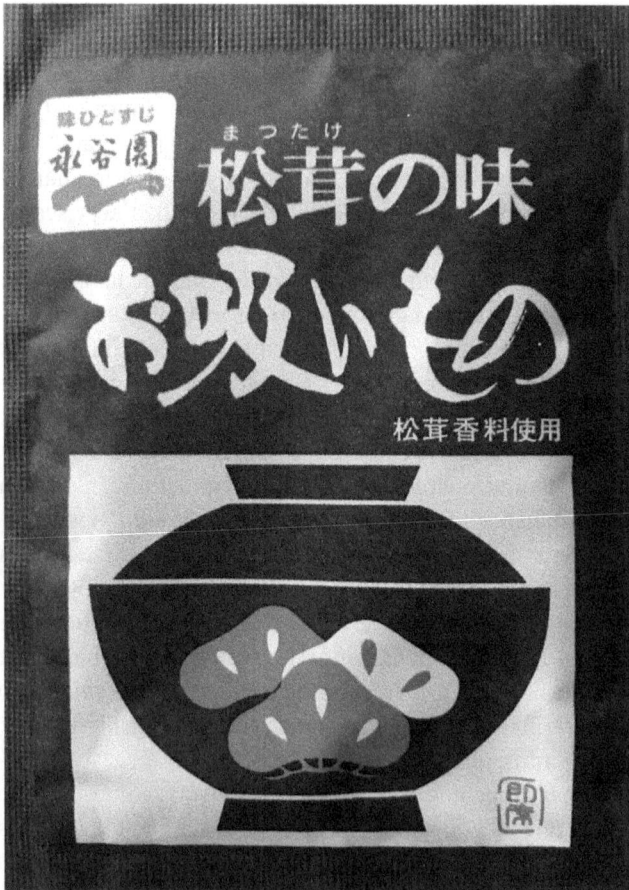

FIGURE 4.1: A package of *matsutake no aji osuimono* (matsutake-flavored soup). Nagatanien Co., Ltd.

time matsutake had already become a luxury item (*zeitaku*), this synthetic soup powder was the original matsutake olfactory experience for many Japanese, while the real matsutake became a mere recollection of its copy.

Unlike the Tsukiji matsutake guru, who unequivocally articulated a tethered indexical relation between the smell of matsutake and its national origin, the elderly people I interviewed had other ways to talk about matsutake: "Ahh, yes, *that* aroma" (*aa ano nioi nee*). So what is "that" aroma? The olfactory power of indexicality is to bring prior sensory experiences to presence. And yet when it comes to matsutake, my consultants paradoxically brought its aroma to presence by pointing to its absence. The matsutake is represented as something that exists over there, not here—then, not now. The charisma of matsutake is thus brought to presence in a series of tropes of deferral, displacement, substitution, and inversion. The aroma of matsutake is located in the past, not the present, and it is to be given to others as a gift, not for one's own consumption—until the future, when it is given to you as a gift. Matsutake is a vehicle with which to circulate money; its value lies not in itself but in its physical adjacency to money. And it is not uncommon to hear people claiming that their very first experience of smelling matsutake comes from synthetic—fake—matsutake soup. The smell of real matsutake is in this way deferred by and inverted with its copy. These tropes make it impossible to get at exactly what matsutake is and what exactly exists in its name. It is as though the unmediated smell of matsutake can be experienced only when it is not there, when the center is empty.

Sense and Language: Can the Nose Speak?

Once language thematizes matsutake as the subject of discourse, and once it aligns itself to what Choy (this volume) calls the "social economy of sense and self," it never reaches the end of its chain of reference, endlessly skirting around and slipping away from its supposed essence. This is, in fact, the general condition under which language refers directly to things in vain (Derrida 1974; Saussure 1966).

Does this mean the triumph of language as an autonomous system independent of reality, or the limited ability of language to represent the thing-in-itself? My consultants would say neither. The matsutake tropes of deferral, displacement, and inversion can be understood as ethics and aesthetics within the realm of linguistic representation, in which my consultants' narratives resist language's dividing practices that create distinctions between the subject and the object, between the human and non-human, and between the sense of smell and the other senses. The tropes of deferral, displacement, and inversion allow my consultants' narratives about matsutake to stop short of being contained by and through familiar cultural idioms and associations, or by political discourses of national identity, and of becoming a symbol of something else, such as 'Japanese-ness', and to remain in the realm of mere possibility and virtuality. When matsutake's smell, wafting in the air as the undifferentiated quality of an ephemeral event (*koto*), enters into language, it refers to matsutake's charisma only by circling around it, as if there were an 'it' at its center. The Japanese nose knows why matsutake is so charismatic—precisely because there is no ultimate 'it' to refer to its smell. But perhaps, inadvertently, this 'it' can be the void that makes it possible for an individual singular moment of "Ahh ... this matsutake smells so good!" to be interchangeable and commensurable with millions of others' moments of "Ahh ... this matsutake smells so good!" This commensurability produces the Japanese nose, which boasts of its ability to discern the unique aroma of the matsutake. The Japanese nose, then, allows even those who have never smelled the matsutake or have never cared about it to say, "It smells good."[3]

There are other mushrooms available in Japan, of course. But it is only matsutake that arouses the nation's (imaginary) nose and troubles its language. It is as though the charisma of matsutake demands reverence from anyone whose (imaginary) nose comes into contact with it. And matsutake whispers to you, "Don't trust language!"

Acknowledgments

Many thanks to my Matsutake Worlds Research Group members, with whom I learned the joy of collaboration. I am also deeply indebted to Mary Bucholtz and Kira Hall for their generous engagement with this chapter. Their comments and suggestions helped me to find my voice here. Needless to say, all errors are mine.

Miyako Inoue is an Associate Professor in the Department of Anthropology at Stanford University, where she teaches linguistic anthropology and the anthropology of Japan. She is currently completing a book manuscript on the social history of 'verbatim' in Japanese. It traces the historical development of the Japanese shorthand technique used in the Diet for its proceedings since the late nineteenth century, and of the stenographic typewriter introduced to the Japanese court for the trial record after World War II. She explores how the politico-semiotic rationality of the stenographic modes of fidelity can be understood as a technology of a particular form of governance, namely, liberal governance. Her publications include a recent article on this topic, "Word for Word: Verbatim as Political Technologies" (*Annual Review of Anthropology*, 2018).

Notes

1. The MWRG consists of Timothy Choy, Lieba Faier, Elaine Gan, Michael Hathaway, Shiho Satsuka, Anna Tsing, and myself.
2. See Chumley and Harkness (2013) for a succinct theoretical discussion of qualisigns.
3. See also Choy (this volume) on the discernment of flavor.

References

Ball, Christopher. 2014. "On Dicentization." *Journal of Linguistic Anthropology* 24 (2): 151–173.

Chumley, Lily Hope, and Nicholas Harkness. 2013. "Introduction: QUALIA." *Anthropological Theory* 13 (1–2): 3–11.

Derrida, Jacques. 1974. *Of Grammatology*. Trans. Gayatri Chakravorty Spivak. Baltimore: Johns Hopkins University Press.

Gal, Susan. 2005. "Language Ideologies Compared: Metaphors of Public/Private." *Journal of Linguistic Anthropology* 15 (1): 23–37.

Gal, Susan. 2013. "Tastes of Talk: Qualia and the Moral Flavor of Signs." *Anthropological Theory* 13 (1–2): 31–48.

Kawai, Suimei. 1912. "Matsutake no rei" [Thank-you note for the gift of matsutake]. In *Shintai shōjo shokanbun* [New styles of girls' letter writing], 363–365. Tokyo: Hakubunkan.

MWRG (Matsutake Worlds Research Group). 2009. "A New Form of Collaboration in Cultural Anthropology: Matsutake Worlds." *American Ethnologist* 36 (2): 380–403.

Nakamura, Shundo. 1930. "Matsutake o okuru bun" [A note to send matsutake with]. In *Tegami no kakikata* [How to write letters], 43–44. Tokyo: Yuhikaku.

Peirce, Charles Sanders. 1978. *Collected Papers of Charles Sanders Peirce*. Vol. 1. Cambridge, MA: Belknap Press of Harvard University Press.

Satsuka, Shiho. 2013. "The Charisma of the Wild Mushroom." *RCC Perspectives* 5: 49–54.

Saussure, Ferdinand de. 1966. *Course in General Linguistics*. Ed. Charles Bally and Albert Sechehaye; trans. Wade Baskin. New York: McGraw-Hill.

Tsing, Anna Lowenhaupt. 2015. *The Mushroom at the End of the World: On the Possibility of Life in Capitalist Ruins*. Princeton, NJ: Princeton University Press.

Watsuji, Tetsuro. 1995. "Kinoko gari" [Mushroom hunting]. In *Watsuji Tetsuro zuihitushû* [Watsuji Tetsuro Collected Essays], ed. Megumi Sakabe, 70–71. Tokyo: Iwanami.

Zee, Jerry C. 2020. "Machine Sky: Social and Terrestrial Engineering in a Chinese Weather System." *American Anthropologist* 122 (2): 9–20.

••:::•••
•••••••
•••••

Chapter 5

SENSING MULTISPECIES ENTANGLEMENTS
Koto as an 'Ontology' of Living

Shiho Satsuka

The manga series *Moyasimon: Tales of Agriculture*, written and drawn by Masayuki Ishikawa (2004–2014), was a big hit in Japan.[1] It tells the story of Tadayasu Sawaki, a first-year student at an agricultural university in Tokyo. Tadayasu is the son of a long-established maker of *moyasimon* (lit., 'agents who burn'), the mold starter that sparks fermentation, which is essential for making traditional condiments, such as soy sauce, soybean paste (*miso*), and Japanese rice liquor (*sake*). The fermentation process requires working with and coordinating micro-organisms such as bacteria, yeast, and fungi. Tadayasu is able to see and communicate with these micro-organisms, a special ability his grandfather notices during Tadayasu's childhood (fig. 5.1). His grandfather sent Tadayasu to study with his old friend, an eccentric professor, hoping that Tadayasu will gain scientific knowledge at his friend's laboratory and further advance his special ability of communicating with micro-organisms. The comic

Notes for this chapter begin on page 107.

FIGURE 5.1: Tadayasu and his grandfather. *Moyasimon 1* (Ishikawa 2009: 4–5)
© Masayuki Ishikawa/Kodansha Ltd.

illustrates Tadayasu's interactions with people and microbes in and around the laboratory at the university. Most of the stories of *Moyasimon* focus on the fungi, yeast, and bacteria that Tadayasu and his friends are dealing with in the laboratory to produce *sake* and other alcoholic drinks. Occasionally, the series depicts other micro-organisms in the agricultural fields and forests, including the matsutake fungus.

It is noteworthy that although Tadayasu can see and listen to the fungi and bacteria, he seldom controls them. These microbes voice their interests loudly among themselves, often teasing Tadayasu, but they are preoccupied with their own business and indifferent to human intentions. Tadayasu and his friends are only able to experiment with the micro-environment that they imagine the microbes will like and to guide them to coordinate in making a good product. In the world of *Moyasimon*, it seems that it is humans who are affected by microbes and who are learning to be attentive to these whimsical and fantastical beings. Unlike conventional anthropomorphic depictions, in which non-human beings are simply used for their cute shapes to tell an allegory of the human world, *Moyasimon* invites us to tune our senses to a world that often evades human eyes, consciousness, and intentions, and yet is simultaneously shaped and experienced by humans and microbes.

Micro-organisms have recently gained specific attention in anthropology and social studies of science as scholars see the possibility of tracing the process of biopolitics shaped by microbiological knowledge (e.g., Lowe 2010; Paxson 2008) and its unintended consequences, challenging conventional analytic scale and boundaries (e.g., Helmreich 2009; Hird 2009; Schrader 2010). Paxson and Helmreich (2014: 167) point out that microbes are "pointers to a biology underdetermined and full of yet-to-be explored possibility." They suggest that "the microbial turn" in biology is a prime site for analyzing changing notions of nature as it "marks the advent of a newly ascendant model of 'nature', one swarming with organismic operations unfolding at scales below everyday human perception, simultaneously independent of, entangled with, enabling of, and sometimes unwinding of human, animal, plant, and fungal biological identity and community" (ibid.: 166). Among various kinds of micro-organisms, mushrooms—fruiting bodies of fungi—have been food for thought for anthropologists and sociologists because of their elusiveness and complex interspecies entanglements. They guide human attention to the world outside the conventional episteme of modern science and capitalist value production (e.g., Ingold 2000; Jasarevic 2015; MWRG 2009; Tsing 2015; Yamin-Pasternak 2008).

In this chapter, I illustrate how people cultivate their sensitivity to mushrooms and, more specifically, how the enigmatic charisma of matsutake guides people to develop their awareness of their multispecies entanglements. The recent discussions of multispecies ethnography point out how the practices of human beings are inseparable from those of animals, plants, and microbes (e.g., Dooren et al. 2016; Haraway 2008; Hustak and Myers 2012; Kirksey 2014; Kirksey and Helmreich 2010; Kohn 2013; Lien 2015). These works challenge the human-nature divide by tracing the interactions between humans and non-humans. Building on these works, I argue that the ways people express their fascination with the fungus further extend the realm of human-nonhuman relationships because the kind of relations that fungus make with humans are elusive and hard to place within conventional frameworks. They indicate the possibility of honing our sensitivities toward the world in which we live with other beings, human and non-human, visible and invisible to human eyes, audible and inaudible to human ears.

In order to explore these questions, I introduce the concept of *koto*, developed by psychiatrist and philosopher Bin Kimura, which refers to events that are experienced temporally in interactions but that evade spatial reification into consciousness. Kimura's notion of *koto*, which I will describe in more detail below, hints at an analytic framework for comprehending entangled life-worlds. It is useful in understanding how people open up their senses to multi-species events. To make this point, I discuss two ethnographic cases in Japan: a grassroots forest revitalization movement for matsutake production led by a microbiologist, and a forest biomass research project conducted in a university laboratory. In both cases, due to the difficulty of isolating matsutake from its complex dynamics with other beings, the scientists have shifted the scale of engagement. In the forest revitalization movement, the microbiologist guided people to transform the entire landscape beyond the matsutake-producing forest. In the laboratory project, the researcher monitored the whole undifferentiated 'mass' of entangled tree and fungus roots even though the usual procedure is to isolate fungus mycelium from tree roots. As I will show below, these practices of scale shifting indicate that the primary concern is not to identify discrete and individual entities, but to focus on the entangled relationship itself. In order to feel the interactions among various beings that are not easily captured by human visual and auditory senses, these scientists developed a sensibility to *koto* happening in the multispecies entanglement and techniques to tune into the coordination among various life forms.

In this sense, *koto* can be conceptualized as an 'ontology' of the satoyama coordination as described by Gan and Tsing (this volume). In this chapter, I call out 'ontology' in quotation marks because the discussions on *koto* indicate traces of the struggle to grasp the very notion from European philosophy and translate it into Japanese: it is "almost the same, but not quite" (Bhabha 1994: 86) as it emerged in the interstitial space of translation.[2] The concept of *koto* itself embodies the generative tension of entanglements in which various knowledge traditions and practices encounter, compete, and grapple with each other. By closely engaging with the idea of *koto*, I also address the politics of knowledge translation by pointing out the uneasiness of adopting 'ontology' to describe the worldviews of people to whom the very notion of ontology—and the distinction between ontology and epistemology—is rather foreign, yet who have been struggling to engage with imported knowledge systems despite the uneven relations of power and knowledge.

The Temporality of *Koto*

In his book *Jikan to Jiko* (Time and Self), Kimura (1982) suggests that the key to honing our sensibilities so that we can see formless things and listen to voiceless beings is to shift our conceptual framework from that of *mono* (things) to that of *koto*, which I tentatively translate as 'events' for lack of a better word in English. But it also could be the 'about-ness' of a thing, that is, all matters that the thing is embedded in and connected to as a happening.[3] Attention to *koto* requires us to imagine time differently. It is necessary to reconceptualize our underlying spatio-temporal assumptions by being aware that the common-sense notion of time—measurable and quantifiable in Newtonian space—is a modern construct. We need to release ourselves from the spell of a modernist conceptualization of the world, which privileges the individual human body as a model unit of ontological existence.

Kimura argues that modern science relies on the ontology of *mono*, a way of conceptualizing the world as a universal space filled with things in which humans try to see things by separating them into individual units and by observing them from a perspectival vantage point. In this framework, *mono* or 'things' occupy space, both external physical space and the internal space of human cognition. Things, with their spatial existence, are mutually exclusive. It is impossible for two things to occupy exactly the same physical space simultaneously. It is also difficult for two things to occupy the internal space of human consciousness at the same time. As soon as our consciousness captures one thing, the previous thing that occupied the space of our cognition should fade away in order for our consciousness to operate without disruption. Things also occupy a certain amount of time. For example, a desk has been occupying a certain amount of time since it was constructed and brought into a room. The temporality of things denotes a time that is measurable and quantifiable; in other words, things are made visible through a scale or length of time. The temporality of *mono* requires spatialized representation.

In contrast, *koto* or 'events' could take place simultaneously without being captured by human consciousness. For example, there are many events happening in our bodies, such as physiological reactions and biochemical activations in our organs and cells. Some of these numerous events and happenings might have strong effects on our consciousness, moods, and activities. But these events and happenings, as *koto*, are elusive. As soon as our consciousness tries to capture *koto*, the *koto* ceases to be *koto* and becomes reified as *mono*, a thing that occupies a spatial topology in our mind. In order for *koto* to remain purely *koto*, it needs to remain in a state of indeterminacy, something that cannot be expressed in a spatial representation. Kimura (1982: 18) writes that "*koto* is always in a fluid and unstable condition similar to elements [in a chemistry experiment] ... *koto* constitutes the present [time] of me without being objectified. *Koto* does not occupy space as *mono* does, yet it occupies my time [without taking the space of my consciousness]."[4] As an indeterminate event, *koto* evades the modernist desire of being captured in a spatial representational scheme. Yet it nonetheless exists and constitutes our presence.

As I will explain below, while Kimura's interest in *koto* 'ontology' derives from his practice as a psychiatrist, its scope is not limited to humans. Kimura (1997) argues that in order to understand subjective experiences of life, it is necessary to separate subjectivity from a particular humanistic construct of the 'mental'. The experience of *koto* should not presuppose an inner mental space, a privileged location for the individual self. Instead, it should be understood as a constant process of emerging events in the organism's encounter with the world. In other words, subjectivity should be located in an *aida* or 'interstice' as an intersubjective phenomenon. This idea of intersubjective phenomenon has larger implications in exploring interspecies relations. Kimura's *koto* 'ontology' reminds us that the generative middle region—constituted by intersubjective experiences that are not fixed into ontological and epistemological categories— is never submerged. *Koto* is constantly happening simultaneously alongside the reified *mono-* or thing-oriented conceptualization of the world. In this sense, *koto* and *mono* are not in opposition, nor are they inseparable. *Mono* emerges as a result of the constant happening of *koto*. A sensibility to *koto* urges us *not* to dwell comfortably in the order of things and with the fixating power of an observing gaze that reifies *mono*.

Koto 'Ontology'

In his dialogue with philosopher Tatsuya Higaki, Kimura explains that his exploration of *koto* was inspired by Heidegger's 'ontological difference', the distinction between beings (all that exists) and Being (the fact or thing of existence) (see Kimura and Higaki 2006: 55, 154). In his pursuit of under-standing patients who claimed that they had difficulty feeling the reality of the world, Kimura became interested in the difference between *koto* and *mono*. His patients stated that they did not feel the existence of an object although their bodily senses functioned and responded to an object, so that they could physi-cally use or avoid objects in order to conduct daily life. Kimura noticed that these patients could capture the object as *mono*, but they might have lost the ability to sense *koto*, whereby the object-being existed as an event, a process of the patients' actions of making relationships with their surroundings (ibid.: 154). He came to realize that what his patients had lost a grip on was not nec-essarily 'reality' (*Realität*), but rather 'actuality' (*Wirklichkeit*), in the sense of a continuous act that is constantly in the process of making relationships and is never reified as an isolated entity (ibid.: 156).

In this exploration, Kimura found that while 'ontological difference' might explain the phenomena of *mono*, it does not fully explicate *koto* due to its human-centered tendency. He argues that *koto* needs to incorporate dynam-ics that go beyond human. To overcome this limitation, Kimura was inspired by the idea of the continuity and differentiation of life as presented by Ger-man physician and physiologist Viktor von Weizsäcker, who suggested that life itself does not die—only individual living beings die (Kimura 2005: 78). Inspired by Heidegger and Weizsäcker, Kimura developed the concept of *biolo-gische Differenz* (ibid.: 78–79). He explains that life itself cannot be an object of recognition, unlike the way each organism can be captured by human cognition. Each organism is able to live as the event (*koto*) of living only on the grounds of composing itself in its surrounding world (*umwelt*). In other words, a subject is in a constant act of emerging in a process of differentiation from life itself while also depending on the relationship it generates with other beings in its surroundings (ibid.: 79). Furthermore, by widely and critically reading the work of French thinkers such as Henri Bergson, Michel Henry and Gilles Deleuze, as well as Japanese philosophers, especially Kitaro Nishida, Kimura formulated his own idea of *koto*. To fully understand *koto*, he urges us to think about the virtuality of life itself—the actuality of emerging living entities and the reality of formed and differentiated life—and the 'timing' of this coordination among various beings that constitute the world. The act of living is a process of this contingent and sometimes violent scramble between life itself and the act of differentiation, which actualizes an individual entity in *koto*, an event.

Kimura's idea resonates with Tim Ingold's (2018) notion of 'ontogeny'. Based on ethnographic studies of the Inuit in the High Arctic, Ingold explains that the Inuit conceptualize the relationship between an individual and others differently from conventional modern understandings. He argues that the word *inuk*, generally translated as 'soul' and from which Inuit is derived, is neither plural nor singular. Rather, 'souls'—or lives—are movements, passed along not by division but rather by multiplication in different individual bodies (ibid.: 159). He writes: "'Life as a whole' ... cannot be reached by any procedure of summation, whether additive or multiplicative. It is never complete, nor is it even on the way to completion, since it advances to no end save its continuation. As the generative potential of a world in becoming, life is always going on, a perpetual origination" (ibid.). Ingold also makes the similar point of continuous generation of life by introducing the concept of 'ground' from the work of a Japanese theatrical artist Tadashi Suzuki, stating that, for Suzuki, the ground is "the very source of emergent difference. It gives rise to the features we see, the formations of the landscape, trees and buildings, even people ... trees and people arise from the earth and from boards, respectively, in an ongoing process of differentiation" (ibid.: 161–162). Ingold posits these ideas developed by Inuit and Japanese as that of 'ontogenesis', becoming different by differentiation from life itself, in contrast to the conceptualization of 'being' that assumes its inert essence originates in its own being. He elaborates the idea of ontogenesis—drawing from Gilbert Simondon's notion of individuation, "the continual 'falling out' of being from becoming"—and writes, "life itself (or soul-life) is a never-ending process of individuation, but critically, the differentiations it engenders are concentrated not at some putative boundary with an outside world, but in its internal resonances" (ibid.: 167).

By contrasting ontogeny and ontology, Ingold (2018) critiques some arguments emerging in the recent debates in the 'ontological turn' in anthropology and related fields, which, he suggests, are preoccupied with a diversity or plurality of worlds. He critiques the very idea of conceptualizing difference as 'diversity' because this framing itself is the product of an atomistic ontology that locates an individuated entity as its starting point. Instead, he takes the knowledges of the Inuit and the Japanese seriously to develop his idea of ontogeny. By translating them via Deleuze and Simondon, he argues that difference should be understood as 'differentiation' from continuing life itself. I align with Ingold's uneasiness in using 'ontology' to gloss over various ways of understanding and engaging with and in the world, since capturing it by the binary set of ontology-epistemology is a very specific European philosophical tradition and rather foreign to many people in the world. If we take the provocation by some authors, such as Viveiros de Castro (2004), Marisol de la Cadena (2015), and Martin Holbraad and Morten Axel Pedersen (2017), to take alternative ways of conceptualizing and engaging with the world seriously, to engage with the world as a kind of 'cosmopolitics' (Stengers 2010), and to challenge the colonization of knowledge (Todd 2016), it is necessary to take a closer look at the way people themselves are struggling with knowledge and practices that shape the idea of 'ontology' and how they are making its incommensurability with the existing knowledge/practice commensurable (as a consequence of European colonial expansion and domination). The uneasiness in the gap of translation is a productive site to notice the dynamics and the generation of ideas, practices, engagements, and power relations operating in constant encounters among various beings. Instead of heroically discovering and saving the alternative or multiple ontologies, I am more interested in learning from the way people grapple with 'ontology' or the hegemonic knowledge system they encounter.[5] In this sense, we can learn from Japanese thinkers' discussions of *koto* as they have emerged in the early twentieth century in the encounter with European philosophy.[6]

Kimura's *koto* 'ontology' can be traced back to the philosopher Kitaro Nishida's ([1927] 1965: 6; italics added) endeavor to translate modern European philosophy, especially Husserl's phenomenological ontology:

> It goes without saying that there are many things to respect and learn from the dazzling development of Western culture that evaluates forms and pays close attention to shapes. However, at the root of Eastern culture that has nurtured our ancestors, I wonder if there is submerged the attitudes *to see the forms of formless things, and to listen to the voices of voiceless beings.* Our hearts ceaselessly seek to [feel] these beings. I would like to give philosophical ground to this demand [of seeing formless things and listening to voiceless beings].[7]

Kimura further explores this endeavor by discussing a pair of verbs. Like the nouns *mono* and *koto*, in Japanese, the verbs *aru* and *iru* indicate two different yet complementary modes of existence. Like both *mono* and *koto*, which can be used to translate 'thing', 'Ding', or 'chose' but locate the subject differently, both *aru* and *iru* can be used to translate 'be', 'sein', or 'être', although the state of being in *aru* and *iru* is quite different. In everyday Japanese, *aru* is used for the existence of non-living things as well as plants and mushrooms, whereas *iru* is used for animals and animated beings. According to Kimura, the existence of every living being should be described with *iru*, as living beings are constantly and actively in the process of differentiating and making relationships with their surrounding beings (Kimura and Higaki 2006: 161).[8] He explains that *iru* takes place in 'contingence' and 'contact' with other beings in the surroundings. Therefore, *iru* connotes the importance of timing. The timing of contact, encounters, and coordination between other beings with tensions in their struggle to live in the surroundings constitutes *koto* (ibid.: 192–194).

Kimura's concept of *koto*, situated in the Japanese intellectual history of struggle in translating 'ontology', helps to explain the popularity of *Moyasimon*. The uniqueness and originality of *Moyasimon* can be seen in the way it scrambles *mono* and *koto*, as well as *aru* and *iru*, in a comical manner. At first glance, it might look like a comic version of an Uexküllian expression of *umwelt*, the life-world experienced by a specific organism (Uexküll [1901] 2010; see also Hathaway, this volume). Informed by recent scientific knowledge, the manga provides detailed explanations of microbes and their characteristics. Yet the point is not to teach readers microbiology, but rather to offer a translation between microbiological knowledge and traditional and vernacular arts of attending to micro-organisms. Tadayasu plays the role of mediating between humans and microbes, and his character also indicates another set of mediations: between technoscience and other forms of knowledge. He has gained the special ability to explicitly see and listen to microbes because his family and ancestors have been working as stewards of microbes and have tacitly cultivated sophisticated arts of noticing micro-organisms. Tadayasu is a mediocre first-year student at the university in the *mono*-centered world, but his capability of noticing the world with a *koto* sensibility exceeds the technical approaches of the senior students and makes the professor's mantra of scientific explanation rather funny and ridiculous.

The difference between Tadayasu and his teacher and colleagues might correspond to two different modes of existence, *aru* and *iru*. For Tadayasu, microbes exist in the mode of *iru*, while his colleagues assume that microbes are *aru* (fig. 5.2). I have noticed that some scientists use *iru* in describing the existence of matsutake mycelium even though it is hard to see with human eyes.

Matsutake's Puzzle

Since ancient times, humans have developed a close relationship with fungi and utilized them in the production of food and liquor. Yet fungi's life experiences have posed a mysterious challenge to human knowledge. The forms and structures of fungi are elusive to human sensory perceptions. In classic taxonomy from Aristotle to Linnaeus, which divided life forms into the two kingdoms of plants and animals, fungi were enigmatic, although considered closer to plants because of their limited mobility. It was only after Robert Whittaker (1969) advocated for a new classification system consisting of five kingdoms that fungi were recategorized as an independent kingdom, neither plant nor animal. Recent research in molecular phylogeny suggests that fungi are actually closer to animals, having speciated from unicellular organisms called choanoflagellates, which have flagella at one end (Kirk et al. 2011). It is also quite recently that humans have become aware of the important role that fungi play in ecological systems. With the development of soil microbiology in the latter half of the twentieth century, fungi have gained attention because they are looked on as the main decomposers of organic matter and minerals, and also as mediators that form various relationships among diverse actors that constitute ecological systems.

Because of their elusiveness, fungi are good navigators to help humans open their sensibilities, and mycorrhizal fungi in particular have fascinated people. Mycorrhizal fungi create a symbiotic relationship with their host trees and form a complex structure called mycorrhiza (lit., 'fungus roots' in Greek), by entangling their hyphae with the roots of host trees. So far, the dynamics of the symbiosis have been hard for human sensory and cognitive systems to fully discern, and thus it is difficult to artificially cultivate mycorrhizal mushrooms. Most expensive wild mushrooms, such as matsutake, truffle, chanterelle, and porcini, are mycorrhizal mushrooms. Unlike saprobic fungi, such as shiitake and meadow mushrooms, which live on dead plants, it is hard to artificially create a living environment for mycorrhizal mushrooms.

Despite century-long efforts at artificial cultivation, Kazuo Suzuki (2005) states that matsutake in particular is still a 'puzzle' for human eyes that have been trained to see forms and appearances. The life cycle of fungi evades conventional representational schemes of time and space. First, it is difficult to identify the individual bodies of mycorrhizal fungi. Because the fungus continues to make clones during its asexual reproduction stage, mycelium with the same DNA continuously extends. Therefore, it is hard to count the number of individual bodies that constitute the mycelial mat, or *shiro*, from which the fruiting bodies of mushrooms are produced. It is further difficult to identify the beginning and the end of a fungus's life cycle. From long years of observations, Japanese scientists argue that a *shiro*'s 'life expectancy', that is, the length of time in which a *shiro* actively produces mushrooms, is about 50 years. But the relationship between the active time span of a *shiro* and the life expectancy of the fungi themselves is not yet known. Some matsutake scientists speculate that the fungi can stay dormant in the soil for many years.

FIGURE 5.2: Tadayasu at the university. *Moyasimon 1* (Ishikawa 2009: 38–39)
© Masayuki Ishikawa/Kodansha Ltd.

HMM, I DON'T KNOW...

I DON'T THINK I'VE GIVEN IT MUCH THOUGHT BEYOND, "WELL, WOULDN'T THAT BE NEAT."

SHOULDN'T *YOU* BE CURIOUS TOO, SENSEI?

Tadayasu Sawaki Protagonist, the son of a *tane-kōji-ya.* Can see microbes for some reason.

ANYHOW, DO YOU MIND IF I HAVE YOU TAKE A LOOK AT SOME STUFF?

UH, I SUPPOSE.... AS LONG AS IT DOESN'T SMELL AS HIDEOUS AS THAT KIVIAK STUFF.

Kei Yūki Longtime friend of Sawaki's. Son of a sake brewer.

...WELL, HOW ABOUT THAT.

THIS IS THE MISO WE MAKE AT OUR PLACE.

!

Haruka Hasegawa Grad student at an Ag school. One of Professor Itsuki's assistants.

3 8

Put simply, matsutake cannot be captured by the ontology of *mono* or things because the beginning to the end of its life and the boundary of its individual body evade the gaze that has been developed to capture things. A fruiting body of matsutake is both *mono* and *koto*, a thing and an event. It is a form temporarily emergent in a visible shape as a result of a series of specific invisible events, the contingent synchronicity of species living in different scales and rhythms of time.

In the *Moyasimon* comics, the microbes' activities, even when unnoticed by Tadayasu, guide us to the *koto* spatio-temporal sensibilities. The fungus *Tricholoma matsutake* is depicted as a builder of a huge underground architecture within the roots of a tree. Matsutake the builder explains to the other microbes in the soil that they have been working on this project for generations; therefore, they call it their Sagrada Familia, referring to Antonio Gaudí's grand architecture project that began in 1882 and continues up to the present day. The builder sarcastically adds that humans refer to their architecture as 'matsutake' only when it appears on the ground and that they are ignorant of the fungi's long-term underground plans and the events they are engaging with. Matsutake laugh scornfully at humans who are blinded by the *mono* ontology.

But it seems to me that some humans may not be as insensible as the insolent matsutake in *Moyasimon* suggests. The enigmatic charisma of the matsutake mushroom leads some scientists to cultivate the ontogenic sense of *koto* under the appearance of *mono*. Below I will present two examples: one is a grassroots citizens' forest revitalization movement led by a matsutake scientist, and the other is a young scholar's doctoral research, which measures forest biomass.

A Forest Revitalization Troop

Due to the challenge of dissecting the detailed mechanisms of symbiosis between matsutake and its host trees, the artificial cultivation of matsutake has not yet been achieved. Therefore, in order to enhance harvests, some matsutake scientists advocate revitalizing an entire forest to a condition that is suitable for matsutake to produce mushrooms (e.g., Ito and Iwase 1997; Yoshimura 2004). Instead of imposing 'ontological cuts' (Barad 2007) on the entangled multi-species symbiosis and focusing on the mechanism and functionality of each species, this group of scientists tries to understand symbiosis in the ecological system as a whole.

The typical niche for red pines and matsutake in central Japan is a satoyama, the secondary forest near human settlements. As Gan and Tsing (this volume) illustrate, traditionally, humans have selectively coppiced and cleared the forest ground in satoyama, using wood, fallen leaves, and grasses for fuel and fertilizer. These human activities have created dry and open forest ground with poor soil nutrition, which is the ideal habitat for matsutake. Because matsutake is a weak competitor among fungi and microbes, if the soil is rich enough to provide food for other species, matsutake cannot thrive. In satoyama, humans, red pines, and matsutake have co-produced a unique environment for living together.[9]

However, the harvest in Japan has decreased drastically since the 1960s (Arioka 1997). Many scholars and commentators argue that the main cause of the decline was the 'fuel revolution', or the introduction of propane gas to rural communities in the 1950s. Rural communities stopped coppicing woods and collecting fallen leaves, relying instead on fossil fuels and chemical fertilizers. The shift in energy infrastructure is an integral part of the rapid and intensive industrialization that the country experienced in the 1960s–1970s. During this period, the population moved from rural agricultural villages to urban centers on a massive scale. Accordingly, much of the satoyama forest was 'abandoned' and left unattended. By the 1990s, many places in the countryside were in 'unhealthy' ecological situations, and matsutake's niche had decreased. Presently, over 95 percent of matsutake consumed in Japan is imported from China, Korea, Canada, the United States, Mexico, Bhutan, Turkey, Morocco, Sweden, and Finland, among other countries. As the critic Toshiyuki Arioka suggests, in Japan "propane ate up matsutake" (ibid.: 264). To many of the satoyama activists and agricultural experts, the decrease of matsutake represents a deplorable situation of agriculture and forestry in post–World War II Japan, an example of heavy dependence on imported food and timber.

Reflecting these concerns, matsutake has recently become an icon of the satoyama forest revitalization movement. One of the most well-known groups is the Matsutakeyama Fukkatsusasetai (Matsutake Forest Revitalization Troop) in Kyoto, led by Fumihiko Yoshimura, a microbial ecologist who has studied matsutake for decades. After retiring from his previous position at a matsutake research center in northeastern Japan, he returned to Kyoto in 2005. He made agreements with some landowners near a suburban community to use their land for his experiment. He invited his friends to work on the unused forests and to return the matsutake to them. Soon their unique practices of revitalizing a matsutake forest caught the attention of media across the nation. A variety of people have joined the ongoing activities, but the majority of the regular participants are retired urban citizens who used to be corporate workers, teachers, or professionals.

About 30 people gather every week and are involved in a variety of tasks. They cut trees, burn the diseased ones, and rake and remove leaves from the forest, all in order to make the environment suitable for matsutake to grow. The challenge is how to get rid of the extensive amount of removed wood, grass, and fallen leaves because there is no place to hold them. While historically they were part of the circulating resources used as fuel and fertilizer, in the contemporary suburban space they are considered waste or 'litter' with no particular everyday use. In order to make use of the grass and leaves, the group began composting them to turn them into fertilizer. Then, in order to use the fertilizer, they started a small vegetable garden, planted tea and fruit trees, and created rice paddies. In order to consume the logs, they built a kiln for pottery making and experimented with making charcoal. In their efforts to get rid of the forest litter, the by-product of conditioning the forest for matsutake, participants became aware of the existence of a variety of living and non-living beings and events happening among them in the forest. The bodily engagement of recreating traditional agrarian landscapes stimulated the participants' *koto* sensibility.

Many senior participants self-mockingly joked during their gatherings that they would not be able witness the return of the matsutake while they were alive. They frequently told me that it would be nice to harvest matsutake for consumption, but the mushroom had become a secondary issue. They learned of the existence of other beings and found joy in feeling connectedness and entanglement with these beings. One regular participant told me that it was worth coming to the gatherings and doing the physically demanding work even if he did not see any matsutake because these activities revitalized not only the forest but his own body and mind. Another participant explained that simply working in the forest and the field was refreshing as she felt reconnected with a variety of beings who share the land. Yet some of the participants gradually showed frustration about the ineffectiveness of their heavy work after nearly 10 years without any signs of mushrooms. Yoshimura stayed calm. He kept telling people that the matsutake fungus exists (*matsutake kin wa iru*) and that they would appear for sure once the forest was conditioned and the timing was right, even though he could not promise when. Then, suddenly, in the fall of 2016, the first two matsutake were spotted. This event further enlivened the group.

Biomass Studies

Many of the matsutake scientists are not necessarily interested in matsutake itself, although they have developed a keen sense of the fungus. While the scientists have utilized the charisma of matsutake to mobilize people or to explain their project and to get research funding, they have seemed to be more interested in understanding the complex relationships among various beings in the forest and analyzing how mycorrhizal fungi work as important actants in forest dynamics.

Miura Shinji[10] was a young postdoctoral fellow researcher when I met him in 2011 in a university laboratory famous for its expertise in separating and propagating mycorrhizal fungus mycelia. In this laboratory, his doctoral work stood out because his study was all about 'mass'. He collected a mass of mycorrhiza (the entangled roots of trees and fungus) from a plot in the university's experimental red pine forest, dried them, and weighed them. On his computer, he recorded the weighted mass on a map of the plot, divided into a grid, to indicate how much mycorrhiza he had found in particular locations. Then he analyzed the correlation between the size of the mycorrhizal biomass and the climate, measured as rainfall, temperature, and sunlight. It seemed that what mattered to him was only the measurable quantities of the mass in general. He did not even distinguish the species of trees or fungi. The focus of his study was the entangled root biomass, and the most important criterion he used in his measurement was whether or not the mass had formed active mycorrhiza. This was determined by finding a typical Y shape of fine root entanglement, which indicates that the mycorrhiza has been formed.

Muira himself was very humble about his work and told me that he planned to identify the species in his future studies. But according to his supervisor, the significance of Miura's doctoral study is that this is a rare and precious study that traces biomass in a conifer forest over multiple years. Instead of the conventional single-year research on biomass, Miura followed the pine-fungus mycorrhizal relationship for six years, utilizing the laboratory's accumulated data on forest biomass from past years as well, and added a temporal dimension to the research. He explained that he wanted to trace the strategies and activities of mycorrhizal fungi in order to understand the role of mycorrhiza in managing the flow of nutrient exchange among varieties of species and the flexible adaptation tactics that trees and fungi are constantly developing. He expected that this study could be the groundwork for a future project on the dynamics of mycorrhizal relations, which are integral to understanding the constant interactions among the variety of species and the movements of things that constitute the seemingly stable forest. Without watching these activities over multiple years, he said that he would not understand what mycorrhizal fungi and host trees intend to do. Even though the individual bodies of fungi are hard to capture with human senses, he still believes there is a way to feel the existence of fungi and their activities by tuning into the rhythm of their way of living. The unit of his research has shifted from the individual bodies, or even a single species population or community, to the entangled mass of relationships itself. The Y-shaped mycorrhiza is a trace of interspecies encounters and coordination.

At first, I thought Miura's research was the quintessential example of work shaped by the *mono*-centered conceptualization of the world. He seemed to be preoccupied with measurable materials and diligently measured mass. Measuring the mass surely relies on the spatial representational scheme of modern science, and his method of tracing forest biomass over a measurable time is shaped by the temporality of things. But I wonder if he has been simultaneously developing a different kind of sensibility toward non-human beings, shaped by the temporality of *koto*. Instead of further dissecting the mycorrhizal structure by identifying the names of the species and by separating the boundaries of each species as *mono*, he seems to let them be as they are, leaving them unidentified and unnamed as traces of *koto*. What appears to be important for his study is not to focus on setting boundaries of the life and death of each species, but to try to feel the rhythms and syncopation in the flow of life that allows for a diversity of beings to exist, even though we do not know what they are. He is keenly aware that these unnamed existences in the undivided entanglement help constitute our presence. Measuring the biomass is his way of communicating with these fantastical creatures, invisible yet surely existing as subjective beings. This sensibility shapes his scientific work and is rooted in his fascination with fungi. As he put it: "The allure of fungi is that they don't show us their presence all the time, but surely they are there." They live actively with their own intentions and rhythms, which may or may not synchronize with our rhythms of human cognition: in doing so, they play an important role in making the world.

When I first visited the laboratory of his doctoral supervisor, Miura and several graduate students took me on a tour to the locations where they frequently went to collect mycorrhizal mushrooms for their experiments. While in the car, they engaged in lively chatter, but once they got out of the car, they quietly walked to the spot and looked around, saying *"konohen ni iru hazu"* (the fungus should be around here). It was before the time that mushrooms appear on the ground, and I could not see any sign of them. I was puzzled and asked, "How do you know that the fungi are here?" A student told me *"nantonaku"* (for some reason). Another student added, "I cannot explain well, but I feel it." I asked further: "Is it because you came to this spot in previous years and you know it by experience?" While the graduate students were looking for words to explain to me, Miura helped the students, saying, "That is a part of it, but somehow [we sense that] the fungus exists [*iru*] here, maybe because of the subtle atmosphere [*fun'iki*]." The students were learning to cultivate the fine ability of sensing the existence of fungus in experimental test tubes in the laboratory and by frequently visiting the fields and watching the professor's and senior students' practices. Even though the fungus did not appear on the ground, they tuned in to *koto* constantly happening in the forest. These tacit techniques of sensing *koto* do not appear in the scientists' published work. Yet they hone their ability to grasp the subtle feeling of events that happen in the forest or with their experimental equipment. While the students and postdoctoral researchers are expected to articulate the events they sense in the language of a *mono*-centered world during their official training, these *koto* sensing techniques are an important part of their practice.[11]

Conclusion

A grassroots forest revitalization movement led by a matsutake scientist and a young scholar's forest biomass research illustrate how the elusive charisma of matsutake leads people to cultivate their 'arts of noticing' (Tsai et al. 2016; Tsing 2015), by developing a sensibility toward the existence of non-human beings and the temporal coordination among various beings. The practices of some matsutake scientists and followers, including the forest revitalization participants, suggest that people are still able to sense the multispecies entanglements by doing scientific work even in the midst of a capitalist society. These cases in Japan raise questions about the clear dichotomies between 'animist' and 'scientific' or 'traditional' and 'modern' knowledge systems. In order to feel the connection with undifferentiated life itself and position oneself in a web of relationships, one does not need to be 'animist' in a traditional sense, imagined to be found only in exotic places separated from the modern scientific world. These practices could be examples of what Deleuze and Guattari (1987) call a 'minor science', that is, knowledge based on attention to contingent and affective interspecies encounters and appreciation of the vitality and movements of various life forms. This is in constant tension with 'royal science' (or 'major science'), which tries to capture the generative power of life by fixing its order and establishing a sense of truth.

Kimura's *koto* 'ontology' is helpful in understanding how people are still able to sense the relational interactions with these beings without relying on intentional semiotic communication. The very process of the development of *koto* 'ontology' itself also illustrates the struggle in translating human-centered and *mono*-dominated ontology in the encounter between European and Japanese thought. *Koto* brings our attention to temporal events that evade spatial reification and conscious captivity. Yet as Kimura (2005) points out, the relationship between life itself and differentiated life is never harmonious or peaceful. It does not allow us to regress into romantic imaginaries of the wholeness of life. The process of differentiation in life is shaped by tensions, frictions, and struggles that are inevitable in every encounter. It provides a hint for developing an analytic framework for understanding entangled life and the politics of knowing the world.

Acknowledgments

Special thanks to the scientists and satoyama revitalization participants who allowed me to join their activities and shared their insights. This chapter has been benefited from the support of the Social Sciences and Humanities Research Council of Canada (file number 435-2016-1191), the Toyota Foundation, and the University of Toronto Connaught Foundation. The chapter evolved from the paper "The Charisma of Mushrooms: The Multiple Rhythms and the Time to Live with Others" presented at the Matsutake Worlds Research Group's panel at the American Anthropological Association in November 2012. This is the result of collaborative fieldwork in Japan with Michael Hathaway and Anna Tsing since 2005 and conversations with Timothy Choy, Lieba Faier, Elaine Gan, Miyako Inoue, Grant Otsuki, and Mei Zhan. Parts of earlier drafts were presented at the University of Chicago's "Re-mediation" workshop (2013), Stanford University's "Matsutake Worlds" workshop (2015), Aarhus University's "Postcolonial Natures" symposium (2015), the Kadokawa Seminar on "Animated Life" at the University of Tokyo (2016), the Technoscience Salon at the University of Toronto (2017), and the University of California Irvine's "Of Soils, Roots and Streams: A Symposium on Ecology and Japan" (2017). I am grateful to the organizers, participants, and audiences, in particular, Anne Allison, Nils Bubandt, David Fedman, Michael Fisch, Mimi Long, Andrew Mathews, Michelle Murphy, Natasha Myers, Shunsuke Nozawa, Johanna Pokorny, Banu Subramaniam, and Heather Swanson. I thank Masayuki Ishikawa and Kodansha for permission to include *Moyasimon* images in this chapter.

Shiho Satsuka is an Associate Professor of Anthropology at the University of Toronto. Her research concerns the politics of knowledge, discourses of nature and science, and cultural practices of capitalism. She is interested in how divergent understandings of nature are produced, circulated, contested, and transformed in translocal interactions shaped by the global expansion of capitalism. She is the author of *Nature in Translation: Japanese Tourism Encounters the Canadian Rockies* (2015) and a co-editor of *The World Multiple: The Quotidian Politics of Knowing and Generating Entangled Worlds* (2018).

Notes

1. Known in Japan as *Moyashimon*, the comics won several awards, including the prestigious Tezuka Osamu Manga Award, the Kodansha Manga Award, and the Soy Source Culture Award in 2008. It was made into an anime series, which had two seasons, and a live-action television series. The first two volumes have been translated into English (Ishikawa 2009, 2010).

2. Casper Jensen and Atsuro Morita (2012) discuss how Japanese anthropologists have developed their interests in 'the ontological turn' in Euro-American anthropology while they have different intellectual trajectories. Jensen's interview with a leading Japanese anthropologist, Naoki Kasuga, also suggests how Kasuga and other Japanese anthropologists have approached 'reality' in their unique way. While they point out the difference between Japanese and Euro-American anthropologists' approach to 'ontology', their discussion is mainly concerned with the explanatory power (i.e., the relevance and effectiveness) of the ontological approach for ethnographic observation and interpretation (see Kasuga and Jensen 2012). In contrast, my main concern here is the politics of knowledge informed by post-colonial and feminist theory. The focus of this chapter is the struggle of people in uneven power relations between different knowledge systems and practices rather than anthropological scientific explanatory power. I thank Martin Holbraad for bringing this to my attention.

3. The *Kenkyusha New Japanese–English Dictionary* translation of *koto* reads "a thing, a matter, an affair, an incident, an event, a question, circumstances, a fact, a case, an accident, trouble, business, a task, a duty, a need, a cause, a reason, an experience." *Koto* is hard to translate into English as it often includes simultaneously a matter and anything related to the matter. Take for example, *matsutake no koto o kangaeru* (I think about matsutake). Here, *koto* serves the function of 'about'. *Matsutake no koto* encompasses both matsutake and anything related to matsutake, which could include, but not be limited to, the surroundings and circumstances, its own characteristics and nature, its relationship with humans and other beings, history, cultural practices, political economy, and so forth.

4. Please note that unless otherwise indicated, all translations from Japanese are my own.

5. See Satsuka (2018) for more about scientists' translation processes of multiple world-making practices. For a discussion concerning the intersection of multispecies ethnography and the ontological turn, see Tsing (2018).

6. See Raud (2002), Marra (2004), and Liu (1995) on the interaction between Heidegger and Kyoto School philosophers.

7. This might sound like a version of the clash of civilizations discourse. However, it should be noted that Nishida wrote this piece when Japanese intellectuals were still struggling to understand, engage with, and translate newly introduced Western scientific and philosophical knowledge, which was overwhelmingly influencing the Japanese intellect and drastically transforming the political/economic systems and the basic framework of society in order to fit to the standard

of the modern Western nation-state. Thinkers such as Nishida, who are labeled Kyoto School philosophers, were exploring the intellectual possibility for a new way of understanding and engaging in the world stimulated by modern Western thought while simultaneously critiquing and appropriating its specific rationality. As historian Christopher Goto-Jones (2008) points out, the political philosophy of the Kyoto School, like that of Nishida, has much potential in offering an alternative understanding of subjectivity, society, and politics, as its scholars tried to 'overcome' the limitations and contradictions of 'modernity' that had fostered Euro-American imperialism. Historically, the Kyoto School's exploration of overcoming Western modernity was appropriated by the ultra-nationalist regime for its oxymoronic and ironic aspiration of 'liberating' Asia from European and American imperialism by establishing its own anti-(Euro-American)imperialist imperialism (Goto-Jones 2009). Thus, the Kyoto School has often been associated with the development of Japanese imperialism. While recognizing this history, one way to learn from the insights of thinkers like Nishida might be to 're-politicize' their thoughts on knowledge politics in order to decolonize our knowledge from both kinds of imperialism—Euro-American and Japanese. Nishida's exploration could be read as similar to Bruno Latour's (1993: 10–11) critique of the 'modern' knowledge-making practice of 'purification' and his attention to the 'hybrid', a series of entangled events and relations that shape the appearance of objects but evade purification and pose challenges to the modernist scientific gaze. Here, 'hybrid' is not the mixture of reified cultural entities of the West and the East. Rather, it is the indeterminacy emerging from the constant process of translation that exists even in the midst of purification efforts in the process of making scientific knowledge.

8. Kimura explains that *aru* can be synonymous with Heidegger's Dasein and *iru* can correspond to Sein, yet Heidegger moves on to argue that, to be more precise, the difference between Sein and Dasein itself constitutes Sein (Kimura and Higaki 2006: 161–162).

9. This process can be described by Haraway's (2008) term 'companion species', which indicates the mutual becoming of humans and other species.

10. This is a pseudonym.

11. The interesting and similar dynamics between affective feeling and reified scientific language are addressed in Wakana Suzuki's (2015) work on the usage of onomatopoeia in a medical science laboratory in Japan.

References

Arioka, Toshiyuki. 1997. *Matsutake*. Tokyo: Hosei University Press.

Barad, Karen. 2007. *Meeting the Universe Halfway: Quantum Physics and the Entanglement of Matter and Meaning*. Durham, NC: Duke University Press.

Bhabha, Homi K. 1994. *The Location of Culture*. New York: Routledge.

Cadena, Marisol de la. 2015. *Earth Beings: Ecologies of Practice across Andean Worlds*. Durham, NC: Duke University Press.

Deleuze, Gilles, and Félix Guattari. 1987. *A Thousand Plateaus: Capitalism and Schizophrenia.* Trans. Brian Massumi. Minneapolis: University of Minnesota Press.

Dooren, Thom van, Eben Kirksey, and Ursula Münster. 2016. "Multispecies Studies: Cultivating Arts of Attentiveness." *Environmental Humanities* 8 (1): 1–23.

Goto-Jones, Christopher. 2008. "The Kyoto School and the History of Political Philosophy: Reconsidering the Methodological Dominance of the Cambridge School." In *Re-politicising the Kyoto School as Philosophy*, ed. Christopher Goto-Jones, 3–25. New York: Routledge.

Goto-Jones, Christopher. 2009. *Modern Japan: A Very Short Introduction.* Oxford: Oxford University Press.

Haraway, Donna. 2008. *When Species Meet.* Minneapolis: University of Minnesota Press.

Helmreich, Stefan. 2009. *Alien Ocean: Anthropological Voyages in Microbial Seas.* Berkeley: University of California Press.

Hird, Myra J. 2009. *The Origins of Sociable Life: Evolution After Science Studies.* New York: Palgrave Macmillan.

Holbraad, Martin, and Morten Axel Pedersen. 2017. *The Ontological Turn: An Anthropological Exposition.* Cambridge: Cambridge University Press.

Hustak, Carla, and Natasha Myers. 2012. "Involutionary Momentum: Affective Ecologies and the Sciences of Plant/Insect Encounters." *differences* 23 (3): 74–118.

Ingold, Tim. 2000. *The Perception of the Environment: Essays on Livelihood, Dwelling and Skill.* London: Routledge.

Ingold, Tim. 2018. "One World Anthropology." *HAU: Journal of Ethnographic Theory* 8 (1–2): 158–171.

Ishikawa, Masayuki. 2004–2014. *Moyasimon.* Vols. 1–13. Tokyo: Kodansha.

Ishikawa, Masayuki. 2009. *Moyasimon 1: Tales of Agriculture.* Trans. Stephen Paul. New York: Del Rey.

Ishikawa, Masayuki. 2010. *Moyasimon 2: Tales of Agriculture.* Trans. Stephen Paul. New York: Del Rey.

Ito, Takeshi, and Koji Iwase. 1997. *Matsutake: Kajuen kankaku de fuyasu sodateru* [Matsutake: Propagate and grow as in orchards]. Tokyo: Nosangyoson Bunka Kyokai.

Jasarevic, Larisa. 2015. "The Thing in a Jar: Mushrooms and Ontological Speculations in Post-Yugoslavia." *Cultural Anthropology* 30 (1): 36–64.

Jensen, Casper Bruun, and Atsuro Morita. 2012. "Anthropology as Critique of Reality: A Japanese Turn." *HAU: Journal of Ethnographic Theory* 2 (2): 358–370.

Kasuga, Naoki, and Casper Bruun Jensen. 2012. "An Interview with Naoki Kasuga." *HAU: Journal of Ethnographic Theory* 2 (2): 389–397.

Kimura, Bin. 1982. *Jikan to jiko* [Time and self]. Tokyo: Chuokoronsha.

Kimura, Bin. 1997. *Kokoro, karada, seimei* [Mind, body, life]. Nagoya: Kawai Bunka Kyoiku Kennkyusho.

Kimura, Bin. 2005. *Kankei to shite no jiko* [Self as relations]. Tokyo: Misuzu Shobo.

Kimura, Bin, and Tatsuya Higaki. 2006. *Seimei to genjitsu: Kimura Bin tono taiwa* [Life and reality: Dialogues with Kimura Bin]. Tokyo: Kawade Shobo Shinsha.

Kirk, P. M., P. F. Cannon, J. C. David, and J. A. Stalpers, eds. 2011. *Ainsworth & Bisby's Dictionary of the Fungi*. 11th ed. Oxon: CABI Publishing.

Kirksey, Eben, ed. 2014. *The Multispecies Salon*. Durham, NC: Duke University.

Kirksey, S. Eben, and Stefan Helmreich. 2010. "The Emergence of Multispecies Ethnography." *Cultural Anthropology* 25 (4): 545–576.

Kohn, Eduardo. 2013. *How Forests Think: Toward an Anthropology beyond the Human*. Berkeley: University of California Press.

Latour, Bruno. 1993. *We Have Never Been Modern*. Trans. Catherine Porter. Cambridge, MA: Harvard University Press.

Lien, Marianne E. 2015. *Becoming Salmon: Aquaculture and the Domestication of a Fish*. Berkeley: University of California Press.

Liu, Lydia H. 1995. *Translingual Practices: Literature, National Culture, and Translated Modernity—China, 1900–1937*. Stanford, CA: Stanford University Press.

Lowe, Celia. 2010. "Viral Clouds: Becoming H5N1 in Indonesia." *Cultural Anthropology* 25 (4): 625–649.

Marra, Michael F. 2004. "On Japanese Things and Words: An Answer to Heidegger's Question." *Philosophy East and West* 54 (4): 555–568.

MWRG (Matsutake Worlds Research Group). 2009. "A New Form of Collaboration in Cultural Anthropology: Matsutake Worlds." *American Ethnologist* 36 (2): 380–403.

Nishida, Kitarō. (1927) 1965. "Hataraku mono kara miru mono e" [From the acting to the seeing]. In *Nishida kitaro zenshu* [Complete work of Kitaro Nishida], vol. 4, 6. Tokyo: Iwanami Shoten.

Omura, Keiichi, Grant Jun Otsuki, Shiho Satsuka, and Atsuro Morita, eds. 2018. *The World Multiple: The Quotidian Politics of Knowing and Generating Entangled Worlds*. New York: Routledge.

Paxson, Heather. 2008. "Post-Pasteurian Cultures: The Microbiopolitics of Raw-Milk Cheese in the United States." *Cultural Anthropology* 23 (1): 15–47.

Paxson, Heather, and Stefan Helmreich. 2014. "The Perils and Promises of Microbial Abundance: Novel Natures and Model Ecosystems, from Artisanal Cheese to Alien Seas." *Social Studies of Science* 44 (2): 165–193.

Raud, Rein. 2002. "Objects and Events: Linguistic and Philosophical Notions of 'Thingness.'" *Asian Philosophy: An International Journal of the Philosophical Traditions of the East* 12 (2): 97–108.

Satsuka, Shiho. 2018. "Translation in the World Multiple." In Omura et al. 2018, 219–232.

Schrader, Astrid. 2010. "Responding to *Pfiesteria piscicida* (the Fish Killer): Phantomatic Ontologies, Indeterminacy, and Responsibility in Toxic Microbiology." *Social Studies of Science* 40 (2): 275–306.

Stengers, Isabelle. 2010. *Cosmopolitics II*. Trans. Robert Bononno. Minneapolis: University of Minnesota Press.

Suzuki, Kazuo. 2005. "Ectomycorrhizal Ecophysiology and the Puzzle of *Tricholoma matsutake*." [In Japanese.] *Journal of the Japanese Forest Society* 87: 90–102.

Suzuki, Wakana. 2015. "The Care of the Cell: Onomatopoeia and Embodiment in a Stem Cell Laboratory." *NatureCulture* 3: 87–105.

Todd, Zoe. 2016. "An Indigenous Feminist's Take on the Ontological Turn: 'Ontology' Is Just Another Word for Colonialism." *Journal of Historical Sociology* 29 (1): 4–22.

Tsai, Yen-Ling, Isabelle Carbonell, Joelle Chevrier, and Anna Lowenhaupt Tsing. 2016. "Golden Snail Opera: The More-than-Human Performance of Friendly Farming on Taiwan's Lanyang Plain." *Cultural Anthropology* 31 (4): 520–544.

Tsing, Anna L. 2015. *The Mushroom at the End of the World: On the Possibility of Life in Capitalist Ruins*. Princeton, NJ: Princeton University Press.

Tsing, Anna L. 2018. "A Multispecies Ontological Turn?" In Omura et al. 2018, 233–247.

Uexküll, Jakob von. (1909) 2010. *A Foray into the Worlds of Animals and Humans with a Theory of Meaning*. Trans. Joseph D. O'Neil. Minneapolis: University of Minnesota Press.

Viveiros de Castro, Eduardo. 2004. "Perspectival Anthropology and the Method of Controlled Equivocation." *Tipití: Journal of the Society for the Anthropology of Lowland South America* 2 (1): 3–22.

Whittaker, R. H. 1969. "New Concepts of Kingdoms of Organisms." *Science* 163: 150–160.

Yamin-Pasternak, Sveta. 2008. "From Disgust to Desire: Changing Attitudes toward Beringian Fungi." *Economic Botany* 62 (3): 214–222.

Yoshimura, Fumihiko. 2004. *Kokomade kita! Matsutake saibai: Matsutakeyama fukkatsu no hasso to gijyutsu* [It has come all the way here! Matsutake cultivation: Ideas and techniques of matsutake forest revitalization]. Tokyo: Toronto.

Chapter 6

HOW THINGS HOLD
A Diagram of Coordination in a Satoyama Forest

Elaine Gan and Anna Tsing

Satoyama is the characteristic village forest of central Japan. Although the word 'satoyama' is recent, adopted by those aiming to revitalize village forests, it refers to woodlands that emerged from a much older arrangement, in which farmers harvested wood for firewood and charcoal and gathered non-timber forest products ranging from dead leaves for green manure to the matsutake mushrooms that form the subject of this book. Matsutake grow in this area

Notes for this chapter begin on page 153.

with Japanese red pines; Japanese red pines grow with satoyama woodlands; satoyama woodlands grow with disturbances created by farmers. This arrangement of more-than-human interdependencies intrigued us. Satoyama woodlands allowed us to think more generally about the concept of 'coordination' as a feature of 'how things hold'.

How has the question of 'how things hold' come to such importance? This question grows out of the turn to the concept of 'assemblage'. In earlier social theory in both ecology and anthropology, 'communities' were the focus of social cohesion. Worried that this concept overemphasized a taken-for-granted social solidarity, scholars in both fields have turned to assemblage, a more loosely imagined grouping. But this opens the question of how entities in an assemblage relate to each other. With growing evidence of symbiosis and interdependence from ecologists and evolutionary biologists (Gilbert and Epel 2015; Margulis 1981), theorists have reimagined the assemblage not as a static set of autonomous elements but as a dynamic process of 'becoming with' (Haraway 2008). The image of the knot has emerged as a way to imagine cohesion without a prior assumption of collective solidarity. Donna Haraway (2003: 6) has described the world as "a knot in motion." Deborah Bird Rose (2012: 136) writes of "embodied knots of multispecies time." Tim Ingold (2015) imagines knots as nodes in meshworks of lifelines. The word 'knots' offers a vivid image of interconnection within the assemblage. How do knots work? This chapter suggests that knots form through attunements in which humans and non-humans can align with each other through timing to make living in common possible. In this way, concepts of assemblage and of coordination require each other. Juxtapositions of beings gain force *as assemblages* when relations of coordination are thick within them.[1]

Coordinations, for us, are temporal rhythms across varied practices that together produce a new capacity or emergence. Many authors use the term to refer to modernist programs of top-down discipline. Glennie and Thrift (1996: 285), for example, define coordination as "the degree to which people's time-space paths are disciplined to smoothly connect with one another's." In contrast, our use of the term hones in on mutual attunements and accommodations. Like Born (2015), we stress improvisation and conjuncture. The coordinations we seek are created when an element of an assemblage finds openings in relations with another element in the assemblage, enabling interchanges where temporal alignment matters. We argue that temporal coordinations bring assemblages into overlapping trajectories of world building. Coordinations make assemblages historically consequential, even as they are made through the frictions and contingencies of assemblages.

Coordination is elusive. It cannot be grasped in a single impression. Because it works across different intervals and rhythms of time, no single snapshot can capture it. Questions of visual and narrative representation arise. How might we parse temporal complexity and render interconnections more expressively? We turned to the diagram as a graphic form that might help us foreground coordination and multispecies assemblage rather than autonomous humans.

A diagram offers a critical description through selectivity and simplicity. Ours draws from a legacy of attempts to illuminate worlds in motion. Paleolithic cave paintings of Chauvet and Lascaux depict animals in superimposed and serial action. Byzantine clocks of al-Jazari used flows of water to mechanize durations. In the early twentieth century, Hermann Minkowski's spacetime diagrams offered a graphical description of four-dimensional relationships between things in motion. Richard Feynman's diagrams of the behavior of subatomic particles offered new formalisms for physicists. Anthropologists' kinship diagrams show the structure of social bonds. In Gilles Deleuze and Félix Guattari's (1987) *A Thousand Plateaus*, the diagram conceptualizes a manifold of socialities, an interplay of machinic and dicursive assemblages that constitutes a contemporary formation. More recently, Mullarkey (2006: 161) has theorized the diagram as a key tool for philosophies of immanence, an iterative method for outlining and exemplifying arrangements rather than explaining relations through transcendental concepts: "Our diagrams will replace conventional words and concepts with lines, arrows, shapes and spatial arrangements." Like Mullarkey's, our diagram works both theory and description. It informs our attempts to show how coordinations are not only critical to but immanent within how things hold.

Our diagram emerges from advocacy for satoyama woodlands. Until the late 1950s, satoyama woodlands were ubiquitous in central Japan. However, as villages were abandoned to the elderly and as village woodlands were converted to plantations or suburban developments, rural areas changed socially and ecologically. Since it began in the 1970s, a vigorous movement to revitalize satoyama has been growing (Tsing 2015). Satoyama advocates taught us how to see woodlands as a set of coordinations. Their guidance stimulated our attempt to theorize coordination through a diagram. We imagine the diagram, then, as neither Japanese nor Western, but rather as one entry into a set of interacting world-making projects crafted separately and together by plants and fungi, satoyama scientists and volunteers, and an anthropologist working with an artist.

Our diagram is a series of pen-and-ink sketches (our 'plates'). Each plate represents an interval of time and segregates particular axes of coordination, while also allowing the reader to watch how those separate axes layer together into structures of continuity and change. Our diagram allows us to visualize coordination as a series of connected events. Avoiding a hyperrealism that renders all details, as well as overgeneralized abstraction, we explore a diagrammatic method through the workings of satoyama. This means that we tell readers quite a bit about satoyama—as well as the satoyama revitalization movement that has instructed us in how to read satoyama. This chapter, however, is not an ethnography of the satoyama or the satoyama revitalization movement—a job better done elsewhere (e.g., Takeuchi et al. 2003). Our descriptions show the encounters through which the coordinations traced in our diagram arise.

Encounters across difference make and emerge from coordinations; the researcher is also a party to such encounters. In acknowledgement of our entanglements, we offer explanations of how we came to diagram coordination. Wedged between the nine plates, we offer instructions on how to read the diagram and notes on how it is made. We describe generative encounters: between matsutake and pine; between satoyama advocates and anthropologists; between critical landscape ecology, visual art, and feminist science studies. We call attention to ways in which incommensurable differences are sometimes able to encourage each other without homogenization. Taken together, we argue, the concepts of coordination, assemblage, and diagram articulate entangled temporalities and histories, showing how socialities emerge.

Encounters

A visitor might first notice the satoyama forest's openness. This forest is not dark but bright and airy. After a rain, a visitor might also notice the fresh smell. What is often identified as an 'earthy' forest smell is actually the aroma of fungi. Following the smell takes even a transient visitor into multispecies worlds—with their riddles of coordination.

October 2005, Kyoto. A prefectural researcher is showing Satsuka, Hathaway, and Tsing how to smell matsutake in the satoyama forest: leaning forward and walking uphill, so that the ground, with its aromas, comes up before us. Our guide has worked hard to keep this forest open so that pine, the host of matsutake, can survive. He explains himself as only one of many makers working in this site, which is neither a garden nor a wild place. Tsing is excited by how much she can smell. Smell itself is a moment of encounter, the beginning of a coordination (see Choy, this volume). The aroma leads Tsing to ponder the combination of plants, fungi, and people that make this forest, which the prefectural researcher has shown her with so much love.

Tsing snaps a picture of the hillside. Later, as they discuss the interplay of oaks, pines, mushrooms, and people in making the forest, she shows the photograph to Gan.[2] It becomes the basis of a collaboration in which the forest itself comes alive as an effect of more-than-human coordination.

Tsing's photograph becomes Gan's window into the forest.

In order to focus on the red pines, Gan traced their outlines with pen and ink. Gan also included the forest floor and the slope of the hillside because they are the basis for the story that will unfold in the next few plates. The line drawing/ tracing that you see is a visual guide to things growing in relation, rather than a portrait of autonomous subjects. The falling leaf on the upper left introduces the second type of tree: oak. Its position breaks the border of the plate to suggest that oak enacts yet another relational temporality beyond the human. The tracing maintains a fidelity to the original scene, while also simplifying it into just two characters—pine and oak—and two temporal qualities—endurance and ephemerality.

When Tsing visited satoyama forests with restoration participants and advocates, her guides showed her the two kinds of trees, pine and oak, that together structure the assemblage. Pollarded and coppiced broadleaf trees keep the forest open and active, while also providing firewood and charcoal. Pines, which grow in the open spaces of peasant use, are important for more extensive burning, for example, in making pottery, salt, and iron—as well as being hosts for matsutake mushrooms. To keep this species assemblage, undergrowth must be kept down and humus removed, originally for green manure, but now just to keep satoyama in place. If the forest floor is not cleared, evergreen broadleaf trees take over, changing the ecosystem. While each satoyama forest is different, every satoyama host pointed to these architectural features and then explained the many forms of creativity, human and non-human, that might emerge within this architecture: shiitake mushrooms can be raised on oak logs; children can be taught to love the outdoors; wildflowers and birds return to the open woodlands; biofuels can be produced; matsutake mushrooms emerge. All these are possible when the basic structure of the satoyama forest is maintained. Satoyama is a more-than-human architecture made through time.

Encounters allow forests to emerge

Deciduous oaks and red pines give the satoyama woodlands their character. Matsutake grows through encounters with pines, forming mycorrhiza—joint organs of fungus and root. The fungus lives off the plant's carbohydrates. The plant benefits from the water and nutrients made possible by the fungus. Forests emerge as assemblages through mycorrhizal connections; trees with mycorrhiza form forests (Curran 1994).

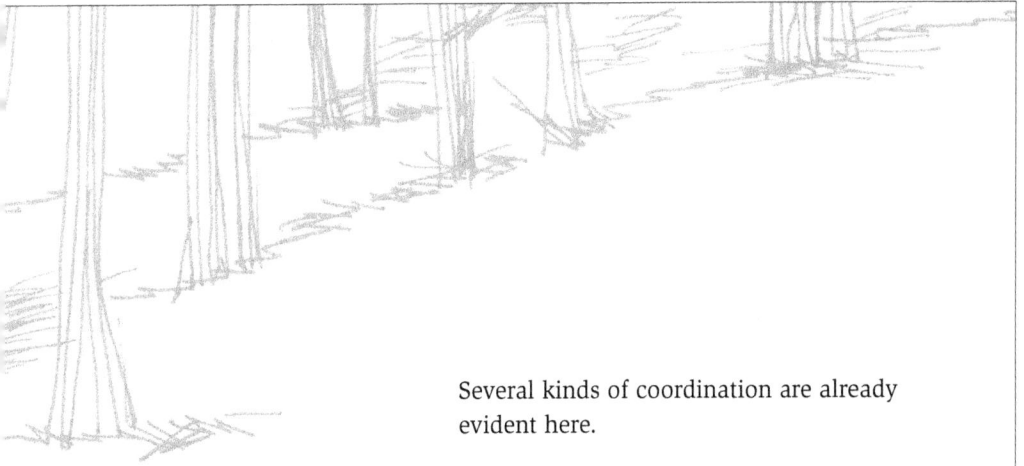

Several kinds of coordination are already evident here.

Evolutionary coordinations
Pines and mycorrhizal fungi evolved together. Pines have developed a particular root structure to suit fungal attachments. Pines and pine fungi are a holobiont, a common evolutionary unit, tied through multispecies coordinations.

Developmental coordinations
Pine seedlings do poorly if they do not encounter mycorrhizal fungi; fungi make it possible for them to thrive. Matsutake, too, need pines.

Successional coordinations
Pines are pioneers, colonizing disturbed places. After some 40 years of pine growth, matsutake may begin to fruit with them.

Masting
Trees and fungi coordinate irregular waves of fruiting.

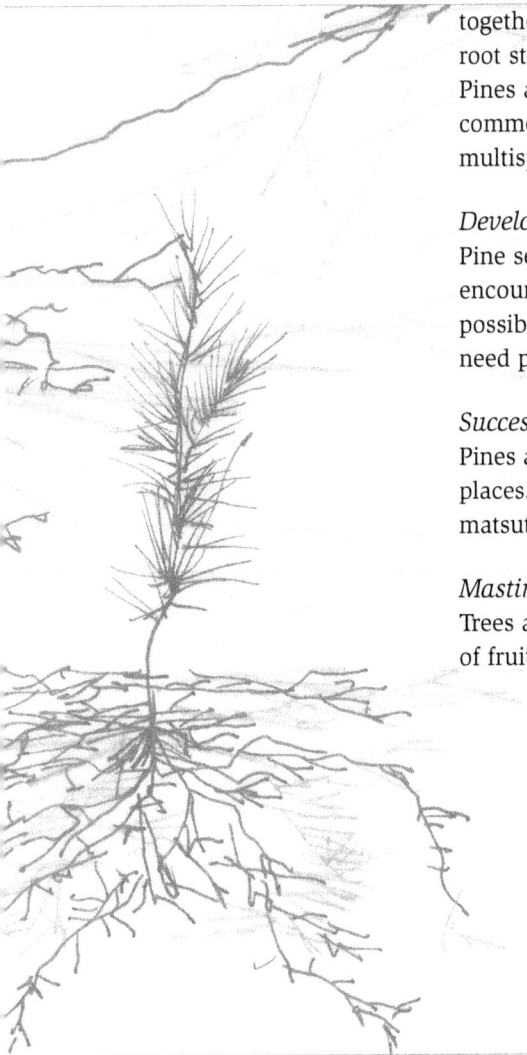

The tracing of pine repeats in this plate, and the oak leaf, carried by wind, continues its journey. A third character appears: mycorrhizal fungi. Because they develop only when pine roots meet fungi underground, mycorrhiza, which are critical to the forest, mostly go unnoticed. Drawing performs an X-ray into the soil.

Here is what matsutake do to make a satoyama forest: they join with pines to colonize open ground and make nutrients available for pines in soils without organic humus. Pines create a forest environment that encourages other plants and animals to move in. Farmers cut back vegetation that might compete with pines; they rake needles and duff, exposing the bare mineral soils pines prefer. Farmers also cut back broadleaf trees, especially oaks, which grow back readily despite the cuts. Regrowing oaks hold hillsides for forests, keeping them from succession to brush or grasslands, despite human disturbance. Each organism offers its own set of temporal cycles and sequences; these temporalities connect in coordinations.

In the plates, we highlight the multiplicity of temporalities by using montage: a filmmaking technique of editing, piecing together, recombining, and overlaying separate shots to convey a novel composition that is based on—and yet analytically distinct from—raw footage from the field (Eisenstein [1949] 1977; MacDougall 1978). Each image is intended to portray its own time as it co-exists with others. Going against graphics software aesthetics of layering different images together and then erasing their seams, our diagram stays close to our observations in order to map and clarify the distinctions between various temporalities.

This plate is one of nine, which together show how a landscape assembles in ways that exceed human management and intention. We use a variety of media: text, field photographs, pen and pencil tracings, and fragmented overlays. We show temporal coordination in four ways that push beyond the limits of written text and digital imagery. First, each of the plates is an act of standing still. It situates a moment from a discrete vantage point. This works like a snapshot, but it does not show just any random moment. Each plate frames an event or an encounter that matters for multispecies livability. Second, across plates, some images repeat to show continuity or change. This works like a flipbook or a film reel, to show how and which encounters come to matter. Encounters socialize place through contingency and emergence. Enduring rhythms are shown through layering, shifts in scale, and changes in transparency. Third, a few plates show discontinuity or a radical change in pattern. Layers are cleared or a landscape altered to show intrusions or indeterminacies that break a sequence. Fourth, each plate opens with a phrase or a sentence that gradually forms a relay across all the plates. Text signals a particular mode of temporality in every scene and shows how they come together historically.

Creating this diagrammatic set of drawings based on hundreds of digital images taken during fieldwork experiments was a method of analysis. Field photographs and videos capture a slice of time, an image of light, an unfolding sequence of actions as they become sensible to a camera or electronic sensor. Multi-dimensional life is flattened into pixels that may be recomposed on a computer at a later date. But our goal was not endless recomposition, the false promise of digital media. We wanted to render particular coordinations and assemblages that unfold differentially and so are not always visible together at the time and place of documentary capture.

As explained above, Tsing was taught to read satoyama for the architectural features that shone through already-ruined versions and now called out for restoration. Satoyama forests might be ideal subjects for a diagrammatic analysis of coordination because satoyama, as enacted in restoration, has the generalized simplicity of a diagram. Its elusiveness is not abolished but rather rendered in a few skillful conceptual strokes, as in an ink brush painting.

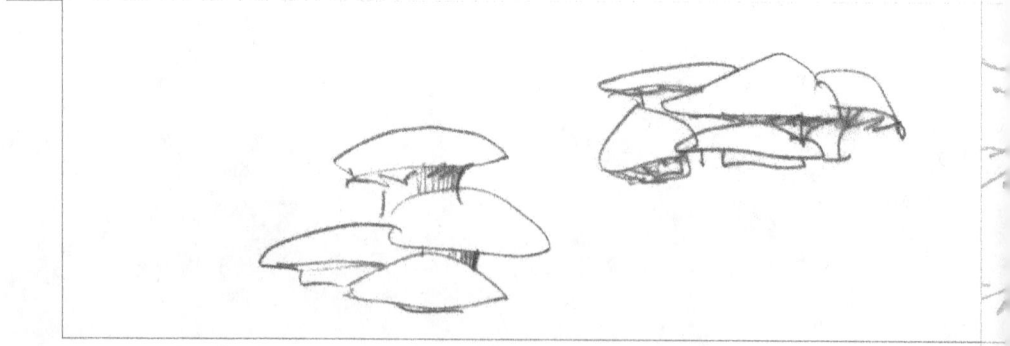

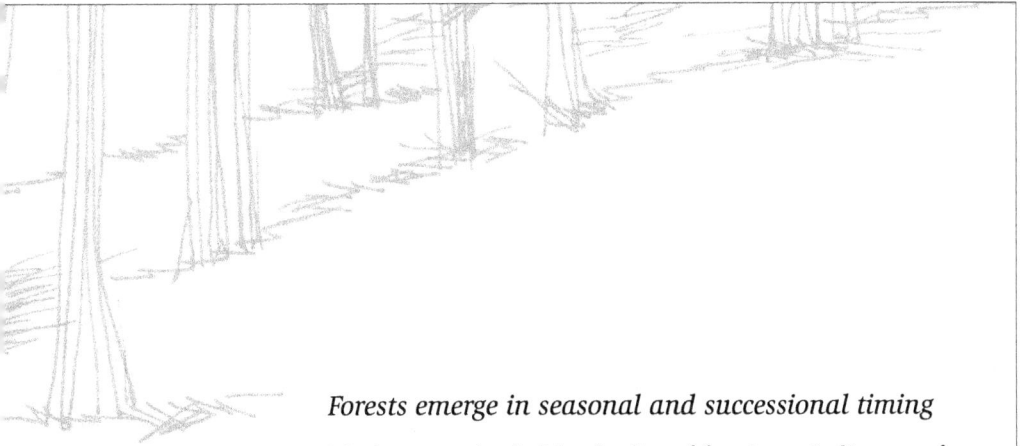

Forests emerge in seasonal and successional timing

Mushrooms, the fruiting bodies of fungi, are indicators of a further axis of coordination: seasonality. As the cool of autumn begins, trees send carbohydrates into their roots. Mycorrhizal fungi respond by fruiting.

Japanese love satoyama woodlands as a theater of Japan's famous four seasons. The open architecture of the woodlands encourages wildflowers in the spring; the deciduous trees turn bright colors in the fall. Animals follow these seasonal changes: birds nest as frogs lay their eggs and hatch in the spring; dragonflies haunt water in high summer as rabbits grow fat in the lush seasonal growth. Foxes chase the rabbits; hawks chase the frogs. Attention to seasonal changes shows the satoyama woodlands as a site for multispecies coordinations—not just tree and fungus, but many varied species.

One way humans join multispecies coordinations is by enjoying their bounty.

Seasons begin to appear. While mushrooms are seasonal, they do not reappear like clockwork. We convey variations within seasonal cycles by drawing the mushrooms in soft clusters, different opacities, and penciled shades. The forest floor is simultaneously changing, and we convey this with a ghosted layering of two of the pine trees. We invite you to find these small variations, which we trace from one plate to the next.

In contrast to diagrams that show how to make a product, our diagram focuses on the temporalities of an assemblage, an open-ended gathering of ways of being. We ask when things meet and start to work together. But just as this redefines the diagram, it also redefines coordination. How might our usage of coordination sit within others' uses of the term?

One well-known treatment of coordination is the actor-network theory of Bruno Latour (2005) and John Law (2008). Actor-network theory reminds us that action emerges from interconnection, which is one kind of coordination. Yet to make the point of emergence clear, Latour purifies his conception of the network from what he calls 'context', that is, those multiple scales of space and time that enter as well as emerge from a chain of associations. Sticking to emergence as it arises within the network, his diagrams are self-consciously 'flat'. As Annemarie Mol (2003: 66) explains it, for Latour "coordination is established or not. There are no distinctive *forms* of coordination." A consideration of temporal coordination requires a richer conceptualization of encounter.

Mol refuses Latourian flatness to ask about coordination within the practices that make up human bodies. How is it, she asks, that medical diagnoses, which differ from a patient's experience of illness, manage to inhabit the patient's body? Mol watches coherence emerge as certain practices are given more authority than others and as ways of enacting bodies are translated into each other's frames. Coordination is not a chain of associations but rather cohabiting ontologies. Still, there are connections: Mol, like Latour, watches coordination to show processes where modernity sees units. When modern units are imagined outside the play of time their processual reworkings are obscured. Mol and Latour move us closer, but timing is not considered as the basis of coordination.

What if instead we began with bodies in motion? This is the perspective advanced by Tim Ingold (1993: 159–160), who shows us how a "taskscape" brings temporality into social life "because people, in the performance of their tasks, *also attend to one another*." We might interpret 'people' in its widest, animist sense as beings; attending is a good place to begin when thinking about temporal coordination. Ingold likens the sociality of taskscapes to orchestral music, in which each musician attends to the conductor and the other musicians. This metaphor opens doors to asking about less perfectly aligned tasks and unruly temporalities. For this, we might turn from a classical orchestra to a baroque fugue in which each instrument maintains a separate melody line. The listener must attend to each line separately as well as to moments of harmony and dissonance as the separate lines cross each other. Separate-and-together listening is required to notice what we call coordination.

In contrast to Ingold's taskscape, our problem of coordination arises in the turbulence and indeterminacy of the assemblage, where specificity and commonality are always taking shape. In contrast to Mol's multiply-enacted body, for which incommensurabilities must be negotiated so that a patient's body can live, our focus on the contingently attuned scene turns us toward evanescent coordinations that hold together precisely because of incommensurabilities. Coherence flickers. Temporality is relational, situating livability for many and diverse bodies. In the multispecies landscape, no single Latourian actor-network shows us enough. Connections both thwart and enable each other in a fugue-like play of life where no resonance ever remains the same. We draw from these scholarly allies to unflatten sociality, even as we follow our own path.

Successional timing includes humans

Humans take part in the coordination; human disturbance forms part of a multispecies portfolio of activities that makes the woodland assemblage possible. Farmers cut wood for firewood and charcoal, which accounts for the woodland's open architecture. Until the advent of fossil fuels, farmers raked leaves and duff for green manure; this removed humus, exposing mineral soils—a gift for pines. Humans cut timber and used fire for shifting cultivation; pines, specialists in disturbed areas, came back. Pines flourished because fungi made it possible for them to find nutrients despite the removal of nutrient-rich humus. Pines and fungi remade human-disturbed places as forests.

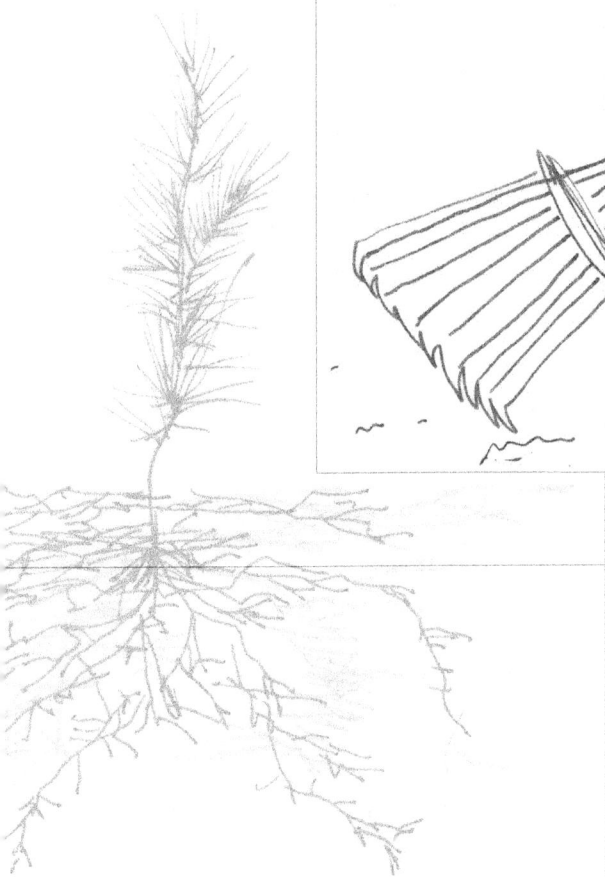

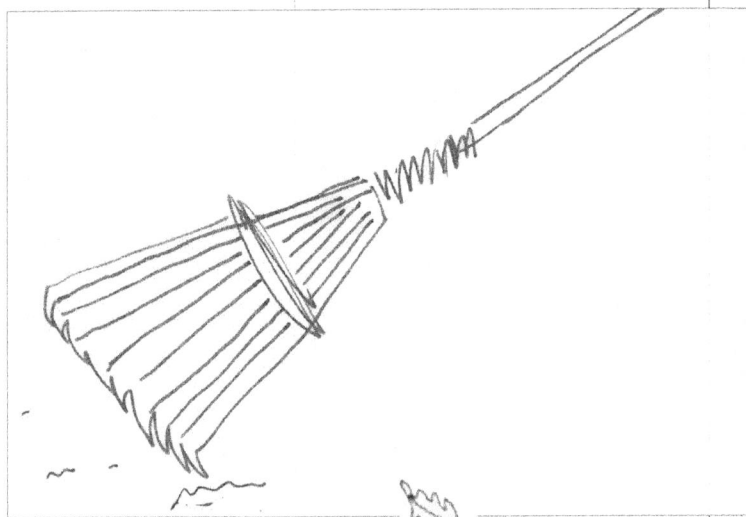

December 2009. Satsuka and Tsing meet with landowners who are trying to revive satoyama. Two partners have taken us to their hillside for the day, where they show us the practices through which they open the forest. Raking is one: the removal of leaves, branches, and duff on the forest floor, once a practice necessary to farming (for the green manure collected), but now a labor of love for the forest. For the moment, they have piled up the plant matter. One partner is hoping they might sell it for biofuel, but they have yet to figure out the business connections. Even without the money, they are committed. This forest was once an active satoyama and a great source for matsutake, and they are determined to restore it.

Raking is a repetitive disturbance that is necessary for the maintenance of satoyama. Gan traced the rake to represent a specific human contribution to forest making. The rake embodies a particular arrangement of local materials: it is not just any rake, nor is it a single rake. The rake, with its own histories and temporalities, merits its own box in this plate. Not all disturbances are acts of destruction. Raking is related to other acts of cultivation—pollarding (cutting branches) and coppicing (cutting trunks)—which appear in the next plate.

But how can we, or the satoyama advocates we follow, claim to know the rhythms of trees and fungi? The key, for us, is to respect the situated nature of knowledge without reducing knowledge to its situatedness. We observe trees and fungi, rather than stopping with advocates' ideas about nature. We also pay attention to the conditions of knowledge production. Three features struck Tsing.

First, description and advocacy of satoyama are part of the same complex of practices. The term 'satoyama' was brought into general use by advocates, whose research defines its characteristics. Our ability to notice the features of satoyama follows advocates as they reassemble relations between city and countryside and between traditional farming and modern conservation. Satoyama advocacy first emerged in central Japan in the 1970s as advocates noted the abandonment of peasant management and the spread of noxious weeds. Young people had moved to the city, and farming had been left to the elderly. As the children of urban migrants grew up, they rediscovered the countryside as a site of advocacy. One early group called itself the Kusakari Jujigun (Grass-Cutting Crusaders), and this name offers a sense of both local challenges and cosmopolitan sources of the movement's appeal. By the twenty-first century, most parts of central Japan had satoyama advocacy groups and rural revitalization projects. Satoyama are not discovered but constituted by these practices. They are simultaneously modern and traditional, and both rural-inspired and urban-inspired.

Second, satoyama advocacy self-consciously mixes science, vernacular enthusiasm, and old farmer knowledge. Elderly farmers are often consultants, but most leaders are urbanites. Some group leaders are scholars and scientists who had begun to worry about the deterioration of the countryside. Other leaders have histories of community organizing. Government foresters play a role in some areas. There are also landowners, some of whom bought land while others inherited family tracts. But their work would not be possible without large groups of volunteers: housewives, students, retired people, and even salaried employees who join in on weekends.

Through this mix of participants, satoyama revitalization brings goals of personal and social revitalization together with scientific inquiry and pedagogy. Some volunteers speak of biodiversity, and some of vitality. Some want rural people to have better incomes. Many are concerned that urban people have lost connection with the countryside and, as a result, have lost something important in themselves. Most have pedagogical goals involving building better relations between people and the countryside around them. Together, these combine what social scientists often imagine as alternative forms of knowledge and practice.

Third, these are both Japanese and not-just Japanese forms of knowledge and practice. They incorporate science, but also aesthetics and philosophy (see Satsuka, this volume). They make connections, extending ties to interlocutors as various as shifting cultivators in Laos and historical ecologists in Sweden. Satoyama advocates build Japanese cultural landscapes, but they do so in conversations that reach across continents.

Taken together, these three characteristics form a basis for an engaged citizen practice that reaches both into and beyond science, Japan, and traditional farming. As much as anthropologists might like this to be true, satoyama advocates are not spokespeople for an autonomous Japanese cosmology. Instead, their agility in moving in and out of these categories, mobilizing translations across varied forms of knowledge and practice, informs their commitments to noticing how pines and oaks and fungi and farmers (as well as scientists and anthropologists) might get along. When living well with difference is taken seriously, the critical role of coordination becomes apparent.

Humans are unintentional actors in the longue durée

Migration opens further axes of coordination. Japan sits in the tracks of two post-glacial migrations: deciduous oaks and pines resemble the flora of the northeast Asian mainland, while evergreen oaks and laurels come from southeast Asia. Peasant disturbance advantaged the former set (Tabata 2001).

Through pollarding and coppicing, farmers helped deciduous oaks in an indirect way. Oaks grew back, remaining stable fixtures in the forest. Meanwhile, farmers removed young evergreens for green manure. Deciduous oaks—with their characteristic companion species, including matsutake—maintained the open architecture of the satoyama woodlands.

The role of human activities became clear when rural Japanese abandoned their farms and woodlands in the 1950s. As soon as pollarding, coppicing, and raking stopped, other species moved into the woodlands, including evergreen oaks and laurels. These species, which remain green all year, produce a dense and dark forest. They create a thick layer of humus, covering mineral soils. Pines fester in the shade and cannot regenerate. Wildflowers disappear, and no leaves change in the fall. Multispecies coordinations change. By the late 1970s, matsutake had become rare.

Thus, too, anthropogenic climate change will be felt through its interruption of multispecies coordinations.

December 2009. Satsuka and Tsing are visiting other satoyama landowners, in this case a husband and wife who are proud to show us the work they have done to make the forest lively. But here we entered a place of old oak pollards, still holding the hillside forest landscape as their own. When encountered in person, the trees' pollarding scars were evident, as was the fact that all the oaks branched at the same height, convenient for humans. Farmers cut back these trees, and they grew into the shapes you see in the sketch.

The shapes of trees are temporal traces. One trace is in the branching itself: in forks and scars we see the results of peasant wood gathering. Another trace follows light. In the sketch, the branches aim upward at steep angles, suggesting that there was no open field around these trees when the branches grew back, but rather shade. A third trace is the mark of time passing. The branches that grew back from the last pollard are now thick and strong, indicating that the pollard was a long time ago. These traces weave a record of the satoyama—and its abandonment.

The trees offer a striking pose, a testimony to relations between people and trees that has lasted many years. In contrast to the appearance of mushrooms, which shows seasonality and indeterminacy, pollarded trees tell a story that lasts the lifetime of the trees (Mathews 2018).

Tree populations also move, and tree migration stories are written in the landscape. Satoyama privileges northeast Asian tree species. When satoyama is abandoned, deciduous oaks are replaced by evergreen oaks and laurels. Amazingly, this is not just a matter of trees: the landscape encouraged by trees draws a selective set of other creatures. Here is Tabata (2001) writing about satoyama ants:

> In Kyoto, *satoyama* woodlands are very rich in ant species because of the mosaic structure of the environment. It is remarkable that rare ant species, especially submerged species, are found in *satoyama* woodlands. They are represented by *Discothyrea sauteri*, *Monomorium triiale*, *Pentastruma canina*, *Epitritus hermerus* and *E. hirashimai*. Boreal species like *Dilchoderus sibiricus* inhabit oak woodlands whereas tropical or subtropical species such as *Epitrutus hexamerus* and *E. hirashimai* are found in bamboo forests. Evergreen broad-leaved forests are also inhabited by tropical or subtropical ant species … Just as the association between fungi and the host plants has been roughly maintained, so has the biological association between particular ant species and vegetation. This association has been maintained at least since the last ice age.

We can only begin to know our many oft-overlooked companions when we learn to think across temporal scales.

Unintentional intrusions matter ...

Sometimes introduced species take off across the landscape, destroying earlier ecologies.

One species capable of invading satoyama woodlands is moso bamboo. Japanese brought this giant bamboo from China 300 years ago in appreciation of its shoots. Every spring, the shoots were harvested; as a result, it never spread. When rural people abandoned their farms in the mid-twentieth century, however, this bamboo spread with a vengeance. Like many grasses, bamboo crowds out other plants. Its underground stems allow it to take over an area without the time lag of reproduction. Bamboo is able to convert satoyama woodlands into dense thickets in just a few years.

Shady conditions weaken pine, making it susceptible to another invasive species, the pine wilt nematode (Suzuki 2004). This tiny worm has caused the rapid decline of pine trees, and thus matsutake, in Japan. Yet this effect has required other coordinations: between New World nematodes and Old World pines, which are vulnerable to their attack; between nematodes and pine sawyer beetles, which carry the worms to the trees; and between human farmers, who no longer cut and rake, and evergreen oaks and laurels, which shade out pines.

Sometimes disarray seems a matter of speed. Bamboo spreads too fast, while Japanese pines are slow to evolve an arrangement with nematodes. But differential speeds impact an assemblage only through coordination: old axes are broken; problematic ones are established; one assemblage disperses, opening up conditions for another.

December 2008. A professor has taken Tsing to the satoyama revitalization project he started at his university. To emphasize the urgency of opening up the forest, he shows her the moso bamboo stands, which have spread wildly. They have grown so thick that other trees cannot survive in their midst. Later, Gan and Tsing chose moso bamboo to make the point about challenges to satoyama because it is so visually dramatic. In the drawing, bamboos form a curtain, blocking everything else.

Here is a photograph of the more ordinary situation, which Tsing took at a different satoyama revitalization project in June 2006. The project leader took Tsing here, too, to show the problem: a forest in which people are no longer participants. The pines are crowded by a host of evergreen broadleaf saplings. There is so much going on that it is difficult to sketch.

Crowding and shading have encouraged the work of pine wilt nematodes, which have killed pines across Japan, changing the landscape. Pine nematodes can wreak destruction only through a finely honed system of coordinations. Consider their transportation: the nematodes are incapable of making their way to trees by themselves. But American nematodes have been able to synchronize their travels with the life cycles of Japanese pine sawyer beetles. They travel by crawling into the breathing tubes of the beetles—but only when the (American) nematodes are in their last larval stage and as the (Asian) beetles become fully adult. The beetle crawls to a new tree and makes a wound to lay its eggs; only then does the nematode safely emerge to eat wood. What careful timing all this requires (Zhao et al. 2013). The delicacy and variety of what we are calling temporalities is illustrated here.

In many satoyama forests across Japan, one sees only dead remnants of pines. The possibilities for matsutake's growth are cut off. But there is more: a long-standing coordination between human and non-human life has been precluded. Livable coordinations can no longer be taken for granted.

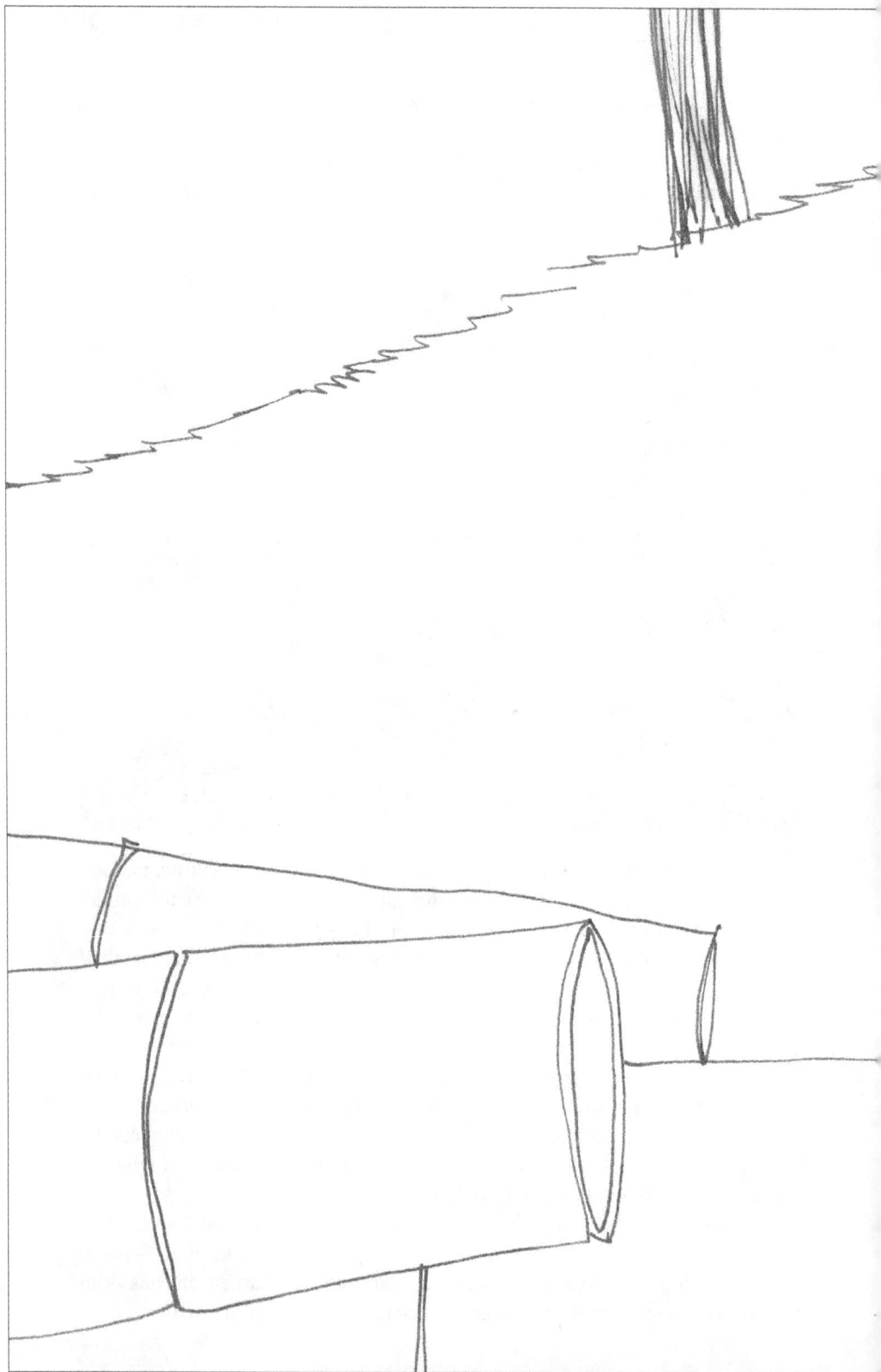

... yet memory is not extinguished

Sometimes assemblages that have been dispersed are able to reassemble. Before the late twentieth century, Japan had many episodes of rapid deforestation, followed by the reappearance of satoyama woodlands. Pine seedlings colonized open ground; oaks followed. This history of satoyama re-establishment has given hope to satoyama advocates, who work to reintroduce human disturbance into Japan's forests to make satoyama woodlands possible again. They clear shady species and remove humus, advantaging pine. They relearn arts of coppice for deciduous oaks.

The word 'satoyama' is more than a descriptive term. Advocates use it to speak of hope for the restoration of particular interspecies arrangements. This is not the same as gardening; it retreats from dreams of mastery. Satoyama advocates do not imagine that they control the assemblage. All they can do is to resume their part in making appropriate disturbances, hoping other species coordinations will follow.

December 2016. Tsing revisits a site for satoyama revitalization. Volunteers have removed the thick brush, allowing red pines to flourish. Amazingly, and for the first time this year, matsutake have emerged. A flag marks one matsutake location. (In the photograph, we focus your attention on the flag and make the rest of the forest recede visually.)

The appearance of matsutake is a small success. It could not be planned. Interactions between matsutake mycelia and red pine roots produce mushrooms; human planning does not. The volunteers make the hillside a good place for pines—and hope and wait. Satoyama advocates take part without expectations of modernist control. Shedding such expectations is key to appreciating both the elusiveness of the mushroom (see Faier, this volume) and the modes of social analysis we see necessary to notice different kinds of coordination.

Modernist dreams have focused attention on only one kind of coordination: disciplined control. Here, elusiveness is banished; every party to the coordination should act predictably. Twentieth-century scholars came to define the term 'coordination' exclusively this way, forgetting that this small subset of coordinations is hardly the only form coordination can take. With only this one meaning in mind, scholars showed that industrialization, urbanization, and the spread of capitalist enterprises required coordinating temporalities. E. P. Thompson's (1967) classic article on industrial time showed the way: workers came to embody the time of the factory. Despite the occurrence of multiple rhythms (Glennie and Thrift 1996), scholars focused on industrial time. Eating changed to fit the factory schedule (Mintz 1986) as human metabolisms were increasingly tethered to the factory. Under this regime, coordination was standardization, and it did not seem elusive—only another instance of discipline.

As our diagram suggests, however, there are other forms of coordination that underlie social life even as their importance is denied by modernist planners. The ecological coordinations of the satoyama are unplanned. Through working with and around each other over a long period of time, pines, matsutake, oaks, and farmers have developed the temporal synchronies that produce the satoyama forest. If we are interested in the dynamics of livability—rather than efficient resource use—coordination in this sense is key: it keeps livable assemblages alive.

Pests and pathogens can be part of an ecological assemblage without destroying it (see Hathaway, this volume). The pine wilt nematode has been successful in killing only so many pines, according to satoyama advocates, because of multiple stresses on pine trees. Furthermore, one disease follows closely on another. The speed and scale of the industrial transfer of pests and pathogens in the last hundred years has been unique. Trees cannot keep up. This is one way that industrial coordinations get in the way of ecological coordinations: the very efficiency of long-distance transfer, along with the scale of industrial plans, has thrown so many pathogens at trees that one species after another has succumbed. Another way that industrial coordinations interfere with forests is to turn them into plantations, that is, monocrops in which planners imagine that only one species is necessary. Tree plantations are an attempt to grow forests without unplanned coordinations. Their gaps in ecological function—and their proclivity to encourage pests and pathogens—make them a poor model for life on earth. Our attention to those coordinations that are unrecognized by plantation planning suggests alternatives.

Encounters layer landscapes

Landscapes are made with multiple projects, human and non-human. Our landscapes are layered with satoyama dreams, mycorrhizal explorations, nematode invasions, autumn matsutake outings, and much more. In coordinations across these enactments, landscapes and assemblages achieve moments of coherence.

In this part of our diagram, we shift from a drawing to a black-and-white photograph. We zoom out from the forest assemblage in previous plates to a wider landscape view.

Satoyama is a landscape that is made over time, through multiple encounters and embodied temporalities. Representations of landscapes too often turn them into backgrounds for human actions. Attending to satoyama enables us to trace landscape's structural features as multiple synchronies, accidental encounters, and indeterminacies.

Although anthropological discussions have not focused on coordinations of this sort, coordinations make it into many ethnographies. In Eduardo Kohn's (2013) *How Forests Think*, for example, an astonishing coordination involving the flight of winged leafcutter ants finds its way into a section on interspecies semiosis. Kohn writes: "The precise moment at which the flight will take place on that day is a response, sedimented over evolutionary time, to what it is that potential predators might, or might not, notice" (ibid.: 80). Once winged ants take off, they become easy prey for bats and birds. Mating ants fly from just before daybreak, when bats are out for only a few more minutes, until right after sunrise, when birds take flight. Bats and birds are least likely to notice the ants within this small yet crucial window in time. Runa, and others interested in ants as food, learn to wait and watch for signs of this elusive event.

Or consider E. E. Evans-Pritchard's (1940) *The Nuer*, which takes us into cascades of coordination. There are daily and seasonal changes, which transform cattle and grass as well as human activities. In the evening, after the day's work is done, a song sung by girls tells of wind blowing *wirawira*, which Evans-Pritchard translates as "the north wind which blows at the time of rich pasture when the cows give plenty of milk" (ibid.: 46–47). Early rains are considered a "season of fatness, for then the grasses germinate, or renew their growth after the long drought, and the cattle can graze on the young shoots to their content" (ibid.: 59). Within routines and rhythms, too, another kind of time arises—fortuitous coincidence: "Nuer, spurred by hunger, leave camp after the first heavy showers to look for giraffe tracks and pursue these animals relentlessly till they overtake them. This is only possible at the time of the first rains when the animals still have to approach camps to drink while their large hooves stick in the moist earth and slow down their movements" (ibid.: 74). Through such coordinations, landscapes are made.

We use the term 'landscape' to refer to material enactments of space and place by many historical actors—human and non-human. The anthropology of landscape was caught for many years in anxieties about the etymology of the word as a European style of painting (Cosgrove 1985); anthropologists could only rehearse differences between these Western ideas of landscape and the ideas of non-Western people (Hirsch and O'Hanlon 1995). More recently, however, there has been a turn to the materiality of landscapes and the enactment of landscape by human and non-human actors (Tilley and Cameron-Daum 2017; Tsing 2015). This turn to materiality opens up a host of new research trajectories, including the search for coordinations.

Sometimes everything changes

Temporal coordinations snap into place and then unravel all the time. Assemblages come and go. Landscapes are made and unmade. Yet ruptures are also possible.

In March 2011, the explosion of the Fukushima Daiichi Nuclear Power Plant changed the satoyama woodlands of northeastern Japan and beyond. Water

and wind carried radioactivity away from the site after the explosion. Just how far one should map sites of rupture is unknown.

As expensive gourmet mushrooms, matsutake have been an icon of satoyama revitalization. Yet mushrooms absorb radiocesium, drawing it into their metabolism. They are a particularly dangerous food in a radioactive landscape. This is one kind of rupture at the heart of the assemblage.

How to represent time out of joint? The black-and-white photograph from the previous plate appears as its inverse and cut apart. The sky turns black; everything is different.

We wrote this chapter in the shadow of the Fukushima disaster. It changed how we imagined how things hold. The disaster was simultaneously biological, geological, and chemical; its effects were immediate and yet will last for many years. It offered a glimpse of a catastrophic coordination, that is, one that abruptly destroyed coordinations that are necessary for livability.

Satoyama, and thus too matsutake, have become more and less common many times in Japanese history. When forests are allowed to grow back without human disturbance, satoyama disappears. When logging creates 'bald mountains' and farmers' practices keep regrowing forests active, satoyama reappears. In these histories, satoyama shows the force of resurgence, the ability of the forest to come back after disturbance, human or otherwise (Tsing 2017). Resurgence is a basic principle of livability. But it is possible to block resurgence—especially within a civilizational matrix in which livability is seen as unimportant in comparison to power and profit. Fukushima radiation is an example of this problem. Fukushima radiation blocks the resurgence of satoyama by the simple fact of making mushrooms toxic, thus undermining satoyama revitalization and the human and non-human coordinations it supports.

A collaboration between photographer Masamichi Kagaya and researcher Satoshi Mori aims to make radioactivity visible through a series of autoradiographs. Below are their images of matsutake from a village 35 kilometers from the Fukushima Daiichi reactor.[3] The irradiated mushrooms in the autoradiograph show contaminated caps and gills.

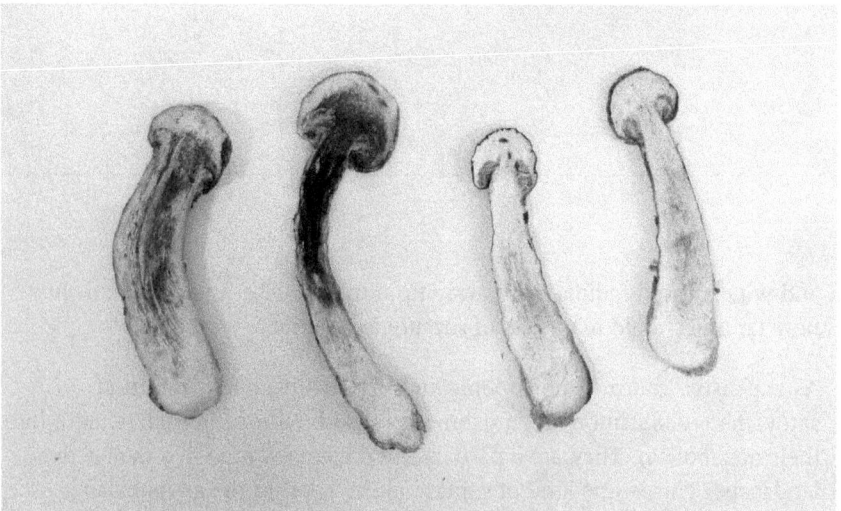

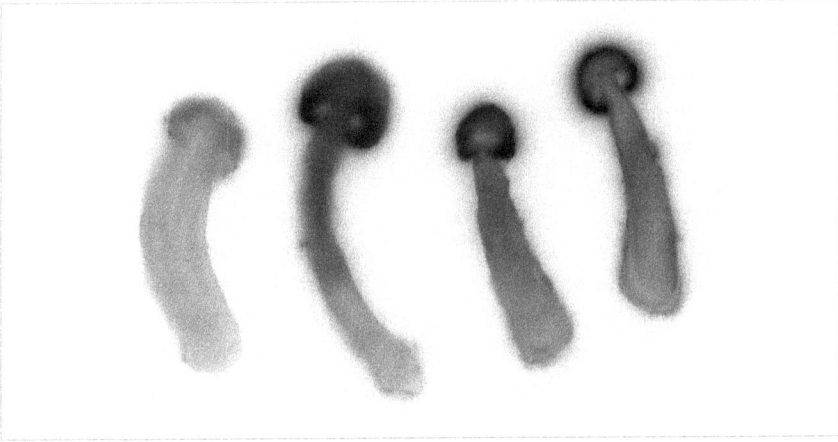

As mushrooms absorb radiocesium, animals (including humans) eat them, transferring radioactivity. Or else the mushrooms decompose, and the radioactivity remains for other life forms to absorb. The coordinations of radioactivity are multiple; it is impossible to measure its temporality as a single line.[4] Still, toxic landscapes will outlast us.

How Things Hold in Precarious Times

Satoyama forests, we argue, show us coordinations, a mode of sociality that does not depend on communication, a common goal, or human-driven webs of significance. Too much popular and scholarly thinking begins with consciousness and communication as the anthropogenic standard with which to rank all living beings. Instead, we show how *timing* can matter to making social lives in common. Through attunements across species difference, as when tree roots and mycorrhizal fungi meet, varied temporalities are brought into coordination, materializing as historically consequential assemblages. Rhythms of life resonate and harness each other, making affordances, too, for others.

It is worth reflecting for a moment on the fact that the question we ask—how things hold—is a significant shift from the classical concerns of both ecology and social science from the last century, when holding together was taken for granted and the important research questions addressed why things fall apart. Ecological disturbance was considered the exception, not the norm, and ties of place and community were imagined as foundational features of the human condition. Today, in comparison, both human and non-human assemblages seem precarious, and the question of whether anything will hold asserts itself with new mystery. Does it make a difference if our landscapes fill up with invasive species and human-disturbance ecologies? In the face of negative pronouncements by ideologues, can neighbors of different races, religions, and nationalities get along? These are the kinds of concerns that have moved social theory to work with open-ended concepts of landscape and assemblage, and to consider how things hold. 'Holding' has become fragile, tentative, and contested. Social theorists suddenly wonder how it works.

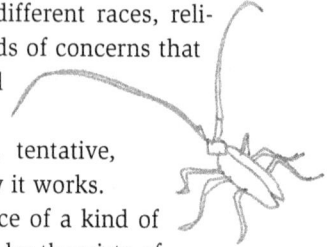

In this chapter, we have argued for the importance of a kind of temporal coordination that was for too long ignored by theorists of modernity: coordination without top-down discipline. We have shown how assemblages emerge through coordination and how coordinations emerge through assemblages. This recursivity happens without common cause or higher authority.

As Gan (2017) argues elsewhere, mapping temporal complexity is key to understanding the Anthropocene. Both the power and terror of industrial civilization can be understood in relation to its more-than-human temporal routines. Coordination is an important tool for this kind of analysis. In tracing coordination in a Holocene ecology, the satoyama forest, this chapter develops a tool for understanding both stability and precarity in the multispecies knots that hold together social assemblages.

Acknowledgments

This chapter is based on the collaborative research of the Matsutake Worlds Research Group, and especially Tsing's joint research with Shiho Satsuka and Michael Hathaway. We are indebted to Fumihiko Yoshimura, Noboru Ishikawa, Daisuke Naito, Heather Swanson, and many others with whom Tsing discussed satoyama. Karen Barad and Michelle Bastian were important interlocutors on multispecies temporalities. This chapter joins many creative experiments between the authors that have taken place from 2012 to 2017. It grew out of a browser-based storytelling project titled "A Fungal Clock: Experiments in Representations of Time," which was initially presented at the annual meeting of the American Anthropological Association in November 2012. The project may be viewed at http://fungalclock.com.

Elaine Gan is an artist-scholar who plays at the intersection of feminist science studies, environmental anthropology, and digital arts and humanities. She teaches at the Center for Experimental Humanities & Social Engagement at New York University. She is a co-editor of *Arts of Living on a Damaged Planet* (2017) and a co-author of "Using Natural History in the Study of Industrial Ruins" (*Journal of Ethnobiology*, 2018).

Anna Tsing is a Professor of Anthropology at the University of California, Santa Cruz, and a Niels Bohr Professor at Aarhus University, where she co-directs Aarhus University Research on the Anthropocene. She is a co-editor of *Arts of Living on a Damaged Planet* (2017) and a co-author of "Using Natural History in the Study of Industrial Ruins" (*Journal of Ethnobiology*, 2018).

Notes

1. For Ingold (2015), correspondence is an alternative to assemblage, which Ingold believes includes too much incoherence. In this, he takes on the question of the best translation for Deleuze and Guattari's (1987) *agencement*. For us, assemblage is a good term because we remain in conversation with its use in landscape ecology. Verran's (2009) discussion of assemblage as an unfolding of heterogeneous agencies with world-making effects also informs our conception.
2. In this chapter, the photographs are by Anna Tsing, and the drawings by Elaine Gan. The autoradiographs were created by Masamichi Kagaya and Satoshi Mori, all rights reserved.
3. See Mori and Kagaya's "Autoradiograph" at http://www.autoradiograph.org.
4. The temporality of radiocesium depends on coordinations. In the laboratory, cesium-137 has a half-life of about 30 years; in the field, however, its ecological half-life is indeterminate. In 2009, Chernobyl's radiocesium level had not changed since the 1986 accident (Madrigal 2009). This is thought to be an effect of mineral flows across layers of soil and organic life.

References

Born, Georgina. 2015. "Making Time: Temporality, History, and the Cultural Object." *New Literary History* 46 (3): 361–386.

Cosgrove, Denis. 1985. "Prospect, Perspective and the Evolution of the Landscape Idea." *Transactions of the Institute of British Geographers* 10 (1): 45–62.

Curran, Lisa. 1994. "The Ecology and Evolution of Mast-Fruiting in Bornean Dipterocaraceae: A General Ectomycorrhizal Theory." PhD diss., Princeton University.

Deleuze, Gilles, and Félix Guattari. 1987. *A Thousand Plateaus: Capitalism and Schizophrenia*. Trans. Brian Massumi. Minneapolis: University of Minnesota Press.

Eisenstein, Sergei. (1949) 1977. "A Dialectic Approach to Film Form." In *Film Form: Essays in Film Theory*, ed. and trans. Jay Leyda, 45–63. New York: Harcourt Publishing.

Evans-Pritchard, E. E. 1940. *The Nuer: A Description of the Modes of Livelihood and Political Institutions of a Nilotic People*. Oxford: Clarendon Press.

Gan, Elaine. 2017. "Timing Rice: An Inquiry into More-Than-Human Temporalities of the Anthropocene." *New Formations* 92: 87–101.

Gilbert, Scott F., and David Epel. 2015. *Ecological Developmental Biology: The Environmental Regulation of Development, Health, and Evolution*. 2nd ed. Sunderland, MA: Sinauer Associates.

Glennie, Paul, and Nigel Thrift. 1996. "Reworking E. P. Thompson's 'Time, Work-Discipline and Industrial Capitalism.'" *Time & Society* 5 (3): 275–299.

Haraway, Donna. 2003. *The Companion Species Manifesto: Dogs, People, and Significant Otherness*. Chicago: Prickly Paradigm Press.

Haraway, Donna. 2008. *When Species Meet*. Minneapolis: University of Minnesota Press.

Hirsch, Eric, and Michael O'Hanlon, eds. 1995. *The Anthropology of Landscape: Perspectives on Place and Space*. Oxford: Oxford University Press.

Ingold, Tim. 1993. "The Temporality of the Landscape." *World Archaeology* 25 (2): 152–174.

Ingold, Tim. 2015. *The Life of Lines*. London: Routledge.

Kohn, Eduardo. 2013. *How Forests Think: Toward an Anthropology Beyond the Human*. Berkeley: University of California Press.

Latour, Bruno. 2005. *Reassembling the Social: An Introduction to Actor-Network-Theory*. Oxford: Oxford University Press.

Law, John. 2008. "Actor Network Theory and Material Semiotics." In *The New Blackwell Companion to Social Theory, Third Edition*, ed. Bryan S. Turner, 141–158. Oxford: Blackwell.

MacDougall, David. 1978. "Ethnographic Film: Failure and Promise." *Annual Review of Anthropology* 7: 405–425.

Madrigal, Alexis. 2009. "Chernobyl Exclusion Zone Radioactive Longer Than Expected." *Wired*, 15 December. https://www.wired.com/2009/12/chernobyl-soil/.

Margulis, Lynn. 1981. *Symbiosis in Cell Evolution: Life and Its Environment on the Early Earth*. San Francisco: W. H. Freeman.

Mathews, Andrew S. 2018. "Landscapes and Throughscapes in Italian Forest Worlds: Thinking Dramatically about the Anthropocene." *Cultural Anthropology* 33 (3): 386–414.

Mintz, Sidney W. 1986. *Sweetness and Power: The Place of Sugar in Modern History*. New York: Penguin Books.

Mol, Annemarie. 2003. *The Body Multiple: Ontology in Medical Practice*. Durham, NC: Duke University Press.

Mullarkey, John. 2006. *Post-Continental Philosophy: An Outline.* London: Continuum.

Rose, Deborah Bird. 2012. "Multispecies Knots of Ethical Time." *Environmental Philosophy* 9 (1): 127–140.

Suzuki, Kazuo. 2004. "Pine Wilt and the Pine Wilt Nematode." In *Encyclopedia of Forest Sciences*, ed. Jeffery Burley, 773–777. Cambridge, MA: Elsevier Press.

Tabata, Hideo. 2001. "The Future Role of *Satoyama* Woodlands in Japanese Society." In *Forest and Civilisations*, ed. Yoshinori Yasuda. New Delhi: Roli Books. http://www001.upp.so-net.ne.jp/ito-hi/satoyama/docs/future.html.

Takeuchi, Kazuhiko, Robert D. Brown, Izumi Washitani, Atsushi Tsunekawa, and Makoto Yokohari, eds. 2003. *Satoyama: the Traditional Rural Landscape of Japan.* Tokyo: Springer-Verlag.

Thompson, E. P. 1967. "Time, Work-Discipline, and Industrial Capitalism." *Past & Present* 38: 56–97.

Tilley, Christopher, and Kate Cameron-Daum. 2017. *An Anthropology of Landscape: The Extraordinary in the Ordinary.* London: UCL Press.

Tsing, Anna. 2015. *The Mushroom at the End of the World: On the Possibility of Life in Capitalist Ruins.* Princeton, NJ: Princeton University Press.

Tsing, Anna. 2017. "A Threat to Holocene Resurgence Is a Threat to Livability." In *The Anthropology of Sustainability: Beyond Development and Progress*, ed. Marc Brightman and Jerome Lewis, 51–65. New York: Palgrave Macmillan.

Verran, Helen. 2009. "On Assemblage." *Journal of Cultural Economy* 2: 1–2, 169–182.

Zhao, Lilin, Shuai Zhang, Wei Wei, Haijun Hao, Bin Zhang, Rebecca Butcher, and Jianghua Sun. 2013. "Chemical Signals Synchronize the Life Cycles of a Plant-Parasitic Nematode and Its Vector Beetle." *Current Biology* 23 (20): 2038–2043.

AFTERWORD
Heeding Headless Thoughts

Eduardo Kohn

This collection of chapters by members of the Matsutake Worlds Research Group (MWRG) is deeply attuned to a certain mode of anthropological thought and practice that I call 'ontological'. Such a mode, perhaps best exemplified by Marilyn Strathern's (1990) *The Gender of the Gift*, consists of a method whereby ethnographic engagement can allow that with which one engages, as well as the mode of attention that such an engagement demands, to be the basis for concept work.[1] From this viewpoint, 'we' do not bring concepts to the world. Rather, we allow ourselves—and this is the methodological challenge—to manifest the concepts that the worlds to which we attend exhibit. This aperture to those thoughts not of our making, and the sensitivity to do epistemic work with them, is what makes this matsutake project 'ontological'.

The kinds of worlds, assemblages, and configurations we think with, then, matter, as they will suggest the concepts that can be part of the engaged modes

Notes for this section begin on page 159.

of being we can 'choose' to cultivate. That this collaborative group 'chose' the form of their engagement—or perhaps more accurately was seduced and maybe even assembled by the matsutake mushroom—points to the potential for a kind of collective wisdom and ethical mode of being for our times.

The ontology of matsutake worlds, then, comprises rich sites for concept work. At the heart of these worlds lie the ephemeral and elusive fruiting bodies of a particular fungus species, which grows best in 'blasted landscapes'. These mushrooms are considered a delicacy in Japan (among other places) and, today, travel through complex commodity chains to end up as non-commodities, as gifts. Their pungent smells and ephemeral freshness serve as signs of the passing of the seasons and remind one to appreciate the evanescent moment of being alive in the face of the certainty of change.

The matsutake are by no means 'domesticated'. What makes matsutake (as well as other prized 'wild' mushrooms such as porcini, chanterelle, morel, and truffle) so unique is that they are mycorrhizal. They live with the living roots of trees. This is also what makes them so hard to cultivate. Our cultivated mushrooms (like champignons, oyster mushroom, and shitake) grow mainly on dead material, which is much easier to replicate in human domestic spaces. By contrast, no one has figured out how to artificially replicate the multispecies forest landscapes that support matsutake.

And yet these spaces are not necessarily 'natural'. Matsutake grow best in highly disturbed landscapes, generally forests whose composition has been greatly affected by humans. For example, they grow well in the 'ruins' of the Oregon clear-cutting lumber industry. And they are intrinsically associated with other-than humans as well (matsutake means 'pine mushroom', in reference to one of its preferred 'allies').

There are other ontological ethnographers of matsutake—Japanese mycologists, for example—and they too do fungally inflected concept work with matsutake. With regard to other studies, the MWRG is wise to point out the limitations of 'Western science' as opposed to 'alter' indigenous modes of thought.[2] A particularly rich example of fungal concept work, as expressed via those humans who attend to it, involves the ways in which Japanese matsutake scientists see these life forms not as 'things' but as 'events'. Perhaps they are saying that when we mistake the mushroom fruiting body for a thing, we miss the fact that what we refer to as an 'it' (the object to be picked, traded, sold, gifted, eaten) is really an elusive and ephemeral sexual activity—an event. The trick for scientists, for mushroom lovers, and for ethnographers of the barely possible is to get in on this act, to promiscuously explode it, and to thereby transform it into what the authors call a 'happening'.

Matsutake are, then, collaborative. They are also 'elusive' (even 'atmospheric'), which becomes another element of concept work. Being elusive, the matsutake refuse conceptual capture. They invite a kind of thinking that does not close off thinking.[3] I should note, however, that closure is sometimes good. I catch a whiff of the all-American, all-you-can-eat buffet here—made infamous by Natasha Schüll (2005). Sometimes closure is a form of doing—the bold and beautiful ink line that captures one image of satoyama and not another. I think too of Ursula Le Guin's (1996) "The Carrier Bag Theory of Fiction." What can 'happen' in that bag is also a result of the selective topology of enclosure that limits what can come in. Form work is concept work too, but it allows things to happen as a result of what cannot.

Different kinds of attunements of multispecies ethnography can suggest very different modes of thought. Many Amazonians, for example, think deeply about apex predators, like jaguars and anacondas, in ways that amplify and heighten 'the self' as a site for concept work. This maps quite comfortably onto the lone anthropological author even as it disrupts him/her by revealing the otherness intrinsic and internal to the self, as well as the 'absential' nature (Deacon 2012) of an unnameable self, forced to navigate an invisible world whose obstacles are the other absences disguised as things.

The MWRG, by contrast, is doing concept work in ecology that has been suggested, seduced, and enabled by the matsutake fungus, which teaches us something altogether different. Fungi have no heads (let alone teeth). Their minds are distributed across their filamentous masses. 'We' headed creatures—dogs, pigs, deer, humans—mistake their sexual organs, their fruiting bodies, for heads. 'We' succumb to 'our' pheromones, which 'they' produce. How can MWRG scholars continue to become authors of 'headless thoughts', the kind of entity that a mushroom is and a jaguar is not? Matsutake demands another practice of the self, another kind of ethics. That is, perhaps, matsutake's self-effacing gift for a planet in ecological trouble.

Eduardo Kohn is an Associate Professor of Anthropology at McGill University and the author of *How Forests Think* (2013), which won the 2014 Gregory Bateson Prize and has been translated into several languages. His research includes human-animal relations and the implications that the ethnographic study of them can have for rethinking anthropology. He also continues to be concerned with discerning guidance from 'thinking' forests.

Notes

1. With regard to concept work, I make little distinction between method and theory. Ethnographic concept work is often as much about how we go about thinking as it is about what we think.
2. Nonetheless, I would insist that there is such a thing as a Western metaphysics, sometimes encapsulated by science, even when it is upended by empirical scientific concept work. I would also insist that there are modes of thought 'alter' to Western metaphysical modes, sometimes expressed by those whom we call 'indigenous'.
3. Members of the MWRG amplify this in a collaborative method that reflects on a kind of thought-work that is not driven by one head or one story.

References

Deacon, Terrence W. 2012. *Incomplete Nature: How Mind Emerged from Matter.* New York: W. W. Norton.

Le Guin, Ursula K. 1996. "The Carrier Bag Theory of Fiction." In *The Ecocriticism Reader: Landmarks in Literary Ecology*, ed. Cheryll Glotfelty and Harold Fromm, 149–154. Athens: University of Georgia Press.

Schüll, Natasha, dir. 2005. *BUFFET: All You Can Eat Las Vegas.* Documentary film, 30 min. https://www.natashadowschull.org/film/.

Strathern, Marilyn. 1990. *The Gender of the Gift: Problems with Women and Problems with Society in Melanesia.* Berkeley: University of California Press.

Index